Clinical Genome
Sequencing

Clinical Genome Sequencing

Psychological Considerations

Edited by

AAD TIBBEN

Clinical Genetics, Leiden University Medical Center, Leiden,
The Netherlands

BARBARA B. BIESECKER

Distinguished Fellow, RTI International, Bethesda MD,
United States

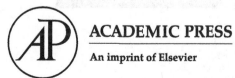

ACADEMIC PRESS

An imprint of Elsevier

Academic Press is an imprint of Elsevier
125 London Wall, London EC2Y 5AS, United Kingdom
525 B Street, Suite 1650, San Diego, CA 92101, United States
50 Hampshire Street, 5th Floor, Cambridge, MA 02139, United States
The Boulevard, Langford Lane, Kidlington, Oxford OX5 1GB, United Kingdom

Notices
Knowledge and best practice in this field are constantly changing. As new research and experience broaden our
understanding, changes in research methods, professional practices, or medical treatment may become necessary.

Practitioners and researchers must always rely on their own experience and knowledge in evaluating and using
any information, methods, compounds, or experiments described herein. In using such information or methods
they should be mindful of their own safety and the safety of others, including parties for whom they have a
professional responsibility.

To the fullest extent of the law, neither the Publisher nor the authors, contributors, or editors, assume any
liability for any injury and/or damage to persons or property as a matter of products liability, negligence or
otherwise, or from any use or operation of any methods, products, instructions, or ideas contained in the material
herein.

British Library Cataloguing-in-Publication Data
A catalogue record for this book is available from the British Library

Library of Congress Cataloging-in-Publication Data
A catalog record for this book is available from the Library of Congress

ISBN: 978-0-12-813335-4

For Information on all Academic Press publications
visit our website at https://www.elsevier.com/books-and-journals

www.elsevier.com • www.bookaid.org

Publisher: Andre Gerhard Wolff
Acquisition Editor: Peter Linsley
Editorial Project Manager: Gabriela Capille
Production Project Manager: Kiruthika Govindaraju
Cover Designer: Greg Harris

Typeset by MPS Limited, Chennai, India

DEDICATION

We dedicate this book to Dr. Seymour Kessler, our teacher and friend who inspired and guided us in our work in the ever-challenging fields of clinical genetics and genetic counseling.

CONTENTS

8. Genetic Counseling and Genomic Sequencing   125
Barbara A. Bernhardt

9. Genome Sequencing in Pediatrics: Ethical Issues   143
Candice Cornelis and Roel H.P. Wouters

10. Genome Sequencing in Prenatal Testing and Screening: Lessons Learned From Broadening the Scope of Prenatal Genetics From Conventional Karyotyping to Whole-Genome Microarray Analysis   157
Sam Riedijk, Karin Diderich, Robert-Jan Galjaard and Gosia Srebniak

LIST OF CONTRIBUTORS

Zaid Afawi
Department of Psychiatry, Erasmus University Medical Center, Rotterdam, The Netherlands;
Department of Family Medicine, Division of Community Health, Faculty of Health Sciences,
Ben-Gurion University of the Negev, Beersheba, Israel

Barbara A. Bernhardt
University of Pennsylvania, Philadelphia, PA, United States

Barbara B. Biesecker
Distinguished Fellow, RTI International, Bethesda, MD, United States

Leslie G. Biesecker
Medical Genomics and Metabolic Genetics Branch, National Human Genome Research
Institute, National Institutes of Health, Bethesda, MD, United States

Christian G. Bouwkamp
Department of Psychiatry, Erasmus University Medical Center, Rotterdam, The Netherlands;
Department of Clinical Genetics, Erasmus University Medical Center, Rotterdam, The
Netherlands

Candice Cornelis
Department of Genetics, University Medical Center Utrecht, Utrecht, The Netherlands;
Ethics Institute, Utrecht University, Utrecht, The Netherlands

Guido de Wert
Faculty of Health, Medicine and the Life Sciences Health, Ethics and Society, Maastricht
University, Maastricht, The Netherlands

Karin Diderich
Clinical Genetics, Erasmus Medical Centre, Rotterdam, The Netherlands

Wybo Dondorp
Faculty of Health, Medicine and the Life Sciences Health, Ethics and Society, Maastricht
University, Maastricht, The Netherlands

Robert-Jan Galjaard
Clinical Genetics, Erasmus Medical Centre, Rotterdam, The Netherlands

Heidi Carmen Howard
Public Health and Caring Sciences, Centre for Research Ethics and Bioethics, Uppsala
University, Uppsala, Sweden

A. Cecile J.W. Janssens
Department of Epidemiology, Rollins School of Public Health, Emory University, Atlanta,
GA, United States

William M.P. Klein
Division of Cancer Control and Population Sciences, National Cancer Institute, Bethesda, MD, United States; National Human Genome Research Institute, Bethesda, MD, United States

Steven A. Kushner
Department of Psychiatry, Erasmus University Medical Center, Rotterdam, The Netherlands

Emilia Niemiec
Public Health and Caring Sciences, Centre for Research Ethics and Bioethics, Uppsala University, Uppsala, Sweden

Sam Riedijk
Clinical Genetics, Erasmus Medical Centre, Rotterdam, The Netherlands

Gosia Srebniak
Clinical Genetics, Erasmus Medical Centre, Rotterdam, The Netherlands

Aad Tibben
Department of Clinical Genetics, Leiden University Medical Centre, Leiden, The Netherlands

Erin Turbitt
National Human Genome Research Institute, Bethesda, MD, United States

Joel Vos
Department of Psychology, University of Roehampton, London, United Kingdom

Marc S. Williams
Genomic Medicine Institute, Geisinger, Danville, PA, United States

Roel H.P. Wouters
Department of Medical Humanities, Julius Center, University Medical Center Utrecht, Utrecht, The Netherlands

CHAPTER 1

Genetic Testing Expanded

Aad Tibben[1] and Barbara B. Biesecker[2]
[1]Department of Clinical Genetics, Leiden University Medical Centre, Leiden, The Netherlands
[2]Distinguished Fellow, RTI International, Bethesda, MD, United States

INTRODUCTION

Humanity has always been aware of heredity in a social and cultural sense, and families have always been aware of diseases that were transferred from one generation to the next. Yet it is only since the 1970s and 1980s that biological knowledge about heredity, more specifically cytogenetics, has allowed clinical genetic diagnosis of newborns with malformations such as Down syndrome, Patau syndrome, and Edwards syndrome. In the late 1980s genetic factors that were associated with diseases with onset later in life were localized and identified using molecular techniques. This allowed confirmation of clinical diagnosis of hereditary diseases such as Huntington disease, and presymptomatic or predictive genetic diagnosis of healthy individuals at risk. Predictive genetic testing resulted in a new category of patients that made the entry into medicine and healthcare, the "future patients," that is, the still healthy individuals who face a future with a high probability of developing a specific disorder. Such a test only informs susceptibility and not whether, when, and how the disease will manifest. A few decades followed of growing interest in predictive testing for Mendelian or single-gene neuropsychiatric and neuromuscular disorders, hereditary forms of cancer, and hereditary cardiovascular disorders. Early identification of carriers of a genetic mutation can be relevant if preventive medical treatment, surveillance options, or lifestyle advice are available. If not, people may be interested in predictive testing as it allows for making informed decisions about their future life perspectives, and also reproduction, with the availability of prenatal genetic diagnosis and preimplantation genetic diagnosis. The development and application of cytogenetics and molecular genetics led to the establishment in the early 1980s of clinical genetics as a new medical specialization. At the same time, psychological research was widely introduced to investigate the impact of genetic testing, and to inform medical-ethical evaluation.

Not only the single-gene disorders but also more common diseases attracted the attention of the basic sciences. The staggeringly rapid developments in the biosciences

Clinical Genome Sequencing
DOI: https://doi.org/10.1016/B978-0-12-813335-4.00001-5

and informatics technology, and the international collaboration of research groups worldwide led to the mapping of the complete human genome in 2001, an endeavor that began in 1990 as the Human Genome Project. With the elucidation of the human genome, research into the function of genes has gained momentum. New technological developments in the field of genome research have greatly expanded research into multifactorial chronic disorders. Genome sequencing has been rapidly integrated into clinical research and has found its way to healthcare practitioners and stakeholders. Exome sequencing and genome sequencing, together described as genome sequencing, are powerful genetic tools that use next-generation sequencing to provide information from nearly all the regions in an individual's genome. The costs of genome sequencing techniques have dramatically declined over the years, which makes them increasingly accessible for large populations. For these reasons, genome sequencing is being used significantly more often in clinical diagnosis and research, thus providing genetic information to patients as well as healthy research participants (Wright et al., 2014). Sequencing is primarily used to identify the etiology of rare diseases but has the potential to become a routine clinical test (Facio et al., 2013; Hitch et al., 2014). In Chapter 2, Genome Sequencing and Individual Responses to Results, Biesecker and Biesecker provide an overview of the clinical uses of genome sequencing, and responses to the return of sequencing information.

High-throughput genome sequencing offers the opportunity to examine large groups of patients and healthy individuals, to learn more about the genetic contribution to disease. Multiple phenotypes are being discovered as well as variant pleiotropy, effective treatments, and reductions in adverse drug effects. Empiric evidence on the psychological impact of the application of these advances is limited but published studies demonstrate broad public interest in the information resulting from genomic testing, and their clinical applications, including return of secondary findings (Bijlsma et al., 2018; Facio et al., 2011, 2013; Gray et al., 2016; Middleton et al., 2016; Shahmirzadi et al., 2014). Middleton et al. (2016) surveyed 6944 people from 75 different countries about their expectations towards sharing secondary findings. Treatability and perceived utility of secondary findings were considered important to 98% of stakeholders personally interested in learning about preventable life-threatening conditions. However, the effects of these developments on the individual and on society require critical consideration from a psychological, societal, and ethical perspective (Biesecker, Burke, Kohane, Plon, & Zimmern, 2012). This book provides a collection of these perspectives to inform healthcare providers and researchers from different fields of medicine and to broaden their frame of reference for genomic testing. We hope that it will be received as a contribution to sound genomic counseling and testing, and future psychological and social science research.

GENOMICS AND SINGLE-GENE TESTING

The development of genomics is in line with and comparable to that of single-gene testing. Yet, there are important differences. Because so many factors are involved in causing multifactorial diseases, they are hard to predict solely on the basis of genomic information. Clinical validity in genome sequencing differs from single-gene testing. The distinction between the etiology of monogenic and multifactorial diseases is not dichotomous, but a spectrum. Huntington disease is caused by one specific expansion of a CAG triplet repeat and can be found at the one end of the spectrum. Diseases such as familial breast cancer can be partially caused by a *BRCA1* pathogenic variant that is incompletely penetrant. Common conditions such as diabetes mellitus, and most psychiatric disorders, develop from complex interactions among the effects of pathogenic gene variants and environmental factors to increase susceptibility and health risks, at the other end of the spectrum. Single-gene testing for both Huntington disease and hereditary breast cancer has clinical utility and validity. This is less clear for multifactorial diseases that necessitate understanding of multidimensional gene—gene interactions and environmental effects.

In analogy to how single-gene testing has changed the medical landscape, the era of genome sequencing will have far-reaching consequences for medical care, and consequently for psychological wellbeing. The breadth of results and frequency of uncertainties inherent to genome sequencing introduce challenges, yet the major issues are ones that we have been addressing in clinical genetics for years. Patients and clients have resources they can draw on to manage the information generated, but attention should be given to identification of those individuals and groups who are most psychologically vulnerable.

THE ERA OF SINGLE TESTING

In the late 1980s and early 1990s, scientific research into the causal factors of hereditary diseases was booming. In 1983, the location was published where chromosome 4 had to be the hereditary cause of Huntington disease (Gusella et al., 1983). It would take another 10 years before the hereditary cause, the expansion of a CAG trinucleotide, was identified (The Huntington's Disease Collaborative Research Group, 1993). The identification of mutations of genes involved in other neurodegenerative diseases and in forms of hereditary cancer and cardiovascular diseases soon followed. Identification of a hereditary factor is important to understand the origin and course of the disease. Only then can a specific treatment be sought or, perhaps, a cure. A byproduct of this research was that identification of a mutation allowed predictive genetic testing, that is the identification of a disease-causing mutation in healthy individuals. This predictive test became possible for the first time in Huntington disease.

The predictive test was a novelty in medicine and heralded a new era in which predictive testing would become available for many other, more or less rare, hereditary diseases that have Mendelian transmission and an onset later in life. Huntington disease had the dubious honor of seceding from the unknown rare diseases and becoming the model that would provide the knowledge and experience to establish predictive testing programs. Since then, predictive testing has become available for several hereditary disorders, including neuropsychiatric diseases such as early-onset Alzheimer disease and frontotemporal dementia, neuromuscular diseases such as myotonic dystrophy, forms of hereditary cancer such as breast and ovarian cancer, colorectal cancer, thyroid gland cancer, von Hippel−Lindau disease and Li−Fraumeni syndrome, and cardiovascular diseases such as familial hypercholesterolemia and cardiomyopathy.

The neurodegenerative and neuromuscular diseases are considered as conditions for which no treatment options are available. Management options for most oncogenetic diseases include regular surveillance, prophylactic surgery, and medical treatment. Surveillance and medical treatment are also options for the cardiogenetic disorders. The far-reaching consequences made predictive testing a matter of social, political, and economic forces. Initially, predictive testing for Huntington disease presented unforeseen and underestimated medical, psychological, and ethical dilemmas. Medically, because we did not have any treatment options to offer. Psychologically, because we did not know how people with the specific perspective of an untreatable disease in the unknown future would build up a full quality of life. Ethically, because we were medical-psychologically empty-handed, and the question arose whether we would do any harm with such a test. Nevertheless, there was great pressure from the worldwide patient advocacy organizations to have such a test.

Introduction and application of the predictive test could not be avoided, and psychological studies were greatly needed to accompany testing programs worldwide (Tibben, 2007). These studies have allowed adjustment of testing protocols and medical-ethical evaluation and have informed support systems and healthcare.

Early studies forecast that most individuals at risk would be tested to make an end to the unbearable uncertainty or obtain certainty to get out of the impasse, and to be able to make important decisions for their future life (Broadstock, Michie, & Marteau, 2000; Duisterhof, Trijsburg, Niermeijer, Roos, & Tibben, 2001; Meiser & Dunn, 2000; Paulsen et al., 2013). We had only vague ideas what a test result might do for people. Indeed, bad news was expected to be dramatic, good news a big relief, which were simple and naive expectations as was demonstrated soon after introduction of the test. With predictive testing for Huntington disease began the odyssey of the predictive test in clinical genetics and other disciplines in medicine, in response to uncertainty or need for certainty. Huntington disease proved to be an important paradigm for all hereditary disorders for which predictive research was possible. In an international collaboration of scientists in Huntington research, clinicians and patient organizations, guidelines for the careful application of the predictive test were published (International Huntington Association & the World Federation of Neurology Research Group on Huntington's Chorea, 1994; MacLeod et al., 2013). Also, recommendations were published for neurologists and psychiatrists on how to address diagnostic, confirmative genetic testing, and predictive genetic testing (Craufurd et al., 2015). Similarly, testing protocols for genetic testing for cancers and cardiovascular disorders were published. A myriad of psychological follow-up studies has been published. These studies have informed us about the importance of sound genetic and psychological counseling, because a genetic test result can have far-reaching consequences in all stages of life (Broadstock et al., 2000; Duisterhof et al., 2001; Meiser & Dunn, 2000; Nance, 2017; Paulsen et al., 2013; Quaid, 2017). Due to the inevitable confrontation with the disease in a parent, children and adolescents could not have a normal childhood, engaging in an intimate relationship could be problematic for young adults, young couples had the fear of starting a family, and once they formed families had to deal with premature failure, financial deterioration, and parents and grandparents had to await how the disease may replicate in their offspring. So, how relevant is what we have learned from these single-gene studies for the genomics era?

Understanding of genetic information is consistently an important part of the process, and individuals should be aware of how the test works, possible outcomes of the test, and opportunities available to manage the health threats, and to participate in research. A vitally important lesson learned from many years of experience with clinical genetic counseling and testing, informed by many psychological studies, is that objective information is only partly relevant for the counselee's thinking, feeling, and behavior (van Dooren et al., 2004; Vos et al., 2008, 2013). This awareness may be even more relevant in genomic counseling. Indeed, it is often the idea that, if people are provided with enough information, they will decide for options that best suit their preferences, personality, or values. This approach is argued as a rational choice or a teaching model method: if people are well informed, they make responsible choices. Clinical genetics and genetic counselors have for years taken into account that people's cognitive skills are not sufficient to making choices. Indeed, social psychologists and behavioral economists have shown for decades that individuals often do not make rational decisions (https://hbr.org/2006/01/a-brief-history-of-decision-making). Patient preferences can be hard to predict, and people often lack the willpower to make planned choices. Moreover, choices are often made unconsciously, not influenced by information. Studies show, for example, that the subjective perception of the results of a genetic test has a greater influence on health behavior and degree of stress than the information provided. People's behavior is influenced by differences in the way they choose or decide to undergo a genetic test. This difference is influenced by the *need for cognition*, the extent to which people tend to think, and find thinking important (Cacioppo & Petty, 1982; Cacioppo, Petty, Feinstein, & Jarvis, 1996). People also differ in the degree of the *need for closure*, that is, to which they want to gain certainty quickly and definitively to eliminate the uncertainty, and in how well they can tolerate uncertainty (Kruglanski, 1989; Webster & Kruglanski, 1994). For example, individuals in a family with a high occurrence of cancer may differ in how they manage the uncertainty of being a carrier of a hereditary mutation. They also differ in dealing with the remaining or residual uncertainty after the outcome of a genetic test, such as the result of a test for a mutation in one of the *BRCA1* or *2* genes, that gives only partial certainty (Vos et al., 2013). These findings suggest the direction parallel psychological studies in the use of genome sequencing may take.

CONSENT TO GENOME SEQUENCING

Adults consent to undergo genome sequencing often in the context of research. Clinical use is also prevalent in cancer genetics, with tumor testing followed by germline testing, if variants that may be in the germline are detected in a tumor. Similarly, panel testing is commonly performed to assess inherited cancer risk and includes many genes in which a variant may be found, increasing the chances for learning a variant of unknown significance. Individual consent is sought in the context of these clinical tests as in research, so that individuals undergoing testing have realistic expectations of the consequences. One consideration is the likelihood of variants of unknown significance that cannot be interpreted. Another is the potential to find a secondary finding that is medically actionable but unrelated to the primary reason for the testing. In pediatrics, parents provide consent for their child to undergo sequencing. They need to understand that they may receive a secondary finding on their child, with the potential for a variant in genes that are expressed in adulthood. The potential overload of genomics information, the complexity of the information, and the limited ability of some patients to sufficiently understand the potential benefits and harms requires prudent consideration of the informed consent process. The advances in genomics challenge traditional conceptions of informed consent (McGuire & Beskow, 2010) and raise questions about what constitutes essential information for participants to understand, and what modes of delivery most promote informed decision-making (Bernhardt et al., 2015). Niemiec and Howard discuss the challenges of consent to genome sequencing in Chapter 3, Consenting Patients to Genome Sequencing.

While a novel context for decision-making, the option to undergo genome sequencing may resemble other (single-gene) genetic testing choices and others in healthcare delivery. As such, research on judgment and decision-making has identified many variables and contexts that may affect decision-making to undergo sequencing, to receive specific results and to act on them, such as notifying relatives of the findings. Klein and others (see Chapter 4: Judgment and Decision-Making in

Genome Sequencing) discuss these decisions and the factors shown to be associated with them including: optimism, information avoidance, tolerance of uncertainty, ambiguity aversion, perceived benefits, perceived risks, perceptions of uncertainty, knowledge, attitudes, and intentions.

Biesecker et al. (2014) examined how research participants perceive the uncertainty of genome sequencing results. Most participants viewed uncertainty as a characteristic of genetic information and described it as developing, unstable, new, and natural. A minority stated that they viewed the uncertainty as questionable, less accurate, limited, or poorly understood. Participants' ideas were related to their beliefs about the nature of genomic knowledge. Those who perceived uncertainty in genomic testing as normal or expected had more optimistic attitudes. Participants who did not expect uncertainty from genomic testing showed more pessimistic attitudes toward receiving results. Biesecker and others discuss the role of uncertainty in genomics (see Chapter 5: Uncertainties in Genome Sequencing).

Another important difference between genome sequencing and single-gene testing concerns secondary findings. Secondary findings are results obtained from genetic testing that do not relate to the primary research or diagnostic reason for the test (Facio et al., 2013). Currently, there is consensus in the United States about the return of analytically valid, clinically actionable findings, that is, findings that yield benefits such as treatment or disease prevention among researchers and clinicians (Kalia et al., 2017), but not about other secondary findings. In Chapter 12, Opportunistic Genomic Screening: Ethical Exploration, De Wert provides an overview of ethical issues and morally relevant characteristics of opportunistic genomic screening and describes the discussion in Europe about how to manage secondary findings.

DIRECT-TO-CONSUMER TESTING

Genome sequencing has become accessible to the masses by direct-to-consumer testing (DCT). Commercial companies already offer genetic susceptibility tests for single-gene diseases but, in 2007, commercial companies appeared on the market with direct-to-consumer genomic testing for a wide range of diseases and other traits. These companies and their online practice received much media attention but were frowned upon by experts and the science was regarded as "questionable" (Bunnik, Janssens, & Schermer, 2014). Interest in DCT has been observed (Boeldt, Schork, Topol, & Bloss, 2015; Hall, Renz, Snapinn, Bowen, & Edwards, 2012), but DCT also raises ethical questions with regard to accuracy, interpretation, and risk−benefit analysis. Janssens expands on DCT in Chapter 6, Direct to Consumer Genome Sequencing, discussing the lack of genomic counseling, follow up, and clinical care, and on how unrealistic consumers' expectations can be.

GENOMICS AND PERSONALIZED MEDICINE

Genomics has the great potential to be used for diagnostic purposes, for pharmacogenomic testing, and for risk assessments for a variety of common disorders or traits and may fine-tune or even transform personalized and precision medicine. Williams (see Chapter 7: Assessing the Psychological Impact of Next-Generation Sequencing Information in the Clinic: An Attempt to Map Terra Incognita?) provides an overarching conceptual framework to inform the implementation of genomics in the clinic. He questions where genome sequencing fits into precision medicine and how realistic its role is. Genomics is emerging in a time where patients are encouraged to have a strong voice with regard to treatment options. All echelons of healthcare need to consider the prerequisites that enable empowerment of patients and clients when applying genome sequencing. One of the fields in medicine that might particularly benefit from genomics is psychiatry, because many genes have been proposed to be involved in the etiology of psychiatric disorders. For most psychic disorders segregating in families, such as bipolar disorder or schizophrenia, no single genes have been identified, which has thwarted the availability of confirmative diagnostic testing or predictive testing. Bouwkamp and others (see Chapter 11: Clinical Genetic Testing and Counseling in Psychiatry) present a conceptual framework within which genomics contributes to psychiatric disorders.

GENETIC COUNSELING FOR SINGLE-GENE TESTING: A PARADIGM FOR GENOMIC TESTING?

Fraser (1974) was the first who articulated a definition of genetic counseling. Genetic counseling aims to identify the needs of the individual or family and uses an empathic client-centered approach (Fraser, 1974). A genetic counselor collects, selects, interprets, and analyzes information, and helps counselees to understand and adapt to the medical, psychological, social, and familial implications of genetic contributions to a specific hereditary disease. A counselor provides an assessment of the chance of disease occurrence or recurrence and educates the counselee on the matter. The counselor promotes informed choices and encourages psychological and behavioral adaptation to the condition or genetic risk. Finally, the counselor facilitates the individual or family to gain access to appropriate healthcare resources. The definition has been adjusted, revised, and expanded given specific hereditary conditions but these key aspects have been maintained across definitions over the years (Biesecker & Peters, 2001; Fraser, 1974; Kessler, 1997, 2001; Resta, 2006; Skirton & Patch, 2009).

Initially, genetic counseling was defined for single-gene testing, guidelines have been adopted by professional organizations and have been endorsed worldwide, several generations of genetic counselors have been trained, and textbooks have been issued that describe the counseling process. In Chapter 8, Genetic Counseling and Genomic Sequencing, Bernhardt addresses the question of whether genetic counseling needs to be redefined in the era of genomics and, consequently, what this means for the professional practice of the genetic counselor. The traditional approach to genetic counseling and genetic testing has been to ensure that an individual has chosen to be tested without coercion and is sufficiently armed with information in order to make an informed decision. No doubt these are important and valuable aims, but whether they are insufficient alone needs to be addressed. Long ago, Kessler (1997) keenly drew attention to the ways genetic counseling can be modeled primarily on a teaching or counseling approach and the challenges of integrating both. Kessler suggests that the goals of each model are distinct: the aim of the teaching model is to have an educated counselee, whereas the counseling approach aims to enhance the self-esteem and autonomy of the counselee. Ongoing, genetic counseling has evolved to a shared decision-making model that combines the best aspects of both models. The practice is not to work through a checklist of what motivates the individual to have the test, and provision of information on the specific disorder and how the test works (MacLeod & Tibben, 2014). Instead, the clinician conducts the consultation using a collaborative approach, partnering with the patient. What has his/her experience been like growing up in a family with a specific hereditary disease? What has led him/her to conclude that a predictive test would be useful at this particular point in their life? What sort of conversations about the disorder have already taken place and with whom? What does (s)he already understand about the symptoms and how the condition is inherited? The use of Socratic questions is particularly helpful in encouraging patients to explore areas they may not yet have considered. A similar approach may be preferred and most effective in the era of genome sequencing.

GENOME SEQUENCING IN PEDIATRICS

Pediatrics is one of the clinical settings where genome sequencing is offered. Diagnosis of rare conditions has advanced, including developmental disabilities coupled with birth abnormalities and complex types of autism due to novel discoveries using genome sequencing. Parents are often seeking to understand the cause of their child's condition, even recognizing that there may be no treatment or cure. Their responses to living with uncertainty may be expected to resemble those seen among parents of children undergoing single-gene testing.

The lack of a prenatal or postnatal diagnosis in a child, using traditional methods, resulted often in uncertain medical care for many years. Genome sequencing technologies allow for far greater interrogation of a child's genome and are therefore useful for reaching a diagnosis in cases of suspected, yet unclarified, genetic conditions. Sapp et al. (2014) examined what kind of secondary findings parents would like to receive about their children with a variety of rare genetic conditions. In their study, all participants wanted to receive variants for treatable and preventable genetic conditions and carrier status. Half of the participants also wanted to receive information about untreatable and unpreventable genetic conditions. Reasons given for obtaining information were that it is parental responsibility to care for their child, that they wanted to obtain the knowledge, that it gave them more control, that it was reassuring to know, and that they wanted to contribute to scientific research. Goldenberg, Dodson, Davis, and Tarini (2014) also examined parental interest in genome sequencing for children with undiagnosed conditions. Most of the parents were interested, or somewhat interested, in genome sequencing for their future newborns. Particularly women, highly educated participants, and participants planning to have a child in the next 5 years, reported interest in genome sequencing. The number of parents who were "definitely interested" in newborn genomic screening declined significantly when they were presented with the possibility that the genomic data could be stored and used in research, suggesting a need for further studies. In Chapter 9, Genome Sequencing in Pediatrics: Ethical Issues, Cornelis and Dondorp describe the special case of use in pediatrics.

GENOME SEQUENCING AND REPRODUCTIVE ISSUES

Schneider et al. (2016) reported two orientations toward genomic preconception carrier screening. The first group consisted of "certain" individuals, who wanted to engage in genomic preconception screening and wanted to obtain all possible results. The second group consisted of "hesitant" individuals, who were cautious about the value of genomic carrier screening and wanted a more limited amount of results. They believed that knowing too much could lead to uncertainty and stress. The most important barrier to engaging in genome screening was the fear and anxiety about what may be discovered. Again, all participants strongly expressed that providing a choice about what results can be obtained was important. In Chapter 10, Genome Sequencing in Prenatal Testing and Screening: Lessons Learned From Broadening the Scope of Prenatal Genetics From Conventional Karyotyping to Whole Genome Microarray Analysis, Riedijk and others report on how they have implemented genome sequencing for prenatal diagnosis in cases of fetal ultrasound anomalies. However, implementation for all indications in prenatal genetic diagnosis is under debate. The main arguments against offering prenatal array testing for all indications are the possibility of detecting copy number variants (CNVs) causing well-described clinically significant anomalies not related to the initial indication (unexpected diagnoses), CNVs associated with a variable expressivity and heterogeneity of clinical features, with an as-yet unquantifiable chance of an abnormal phenotype if found prenatally (so-called susceptibility loci for neurodevelopmental disorders); and variants of unknown clinical significance. These outcomes raise questions about which genomic microarray results should be reported to pregnant couples and whether they should be offered a choice regarding which possible array outcomes they wish to be informed of.

PSYCHOLOGICAL STUDIES IN GENOMICS

Given the far-reaching impact of genome sequencing, the psychological and social challenges surrounding its introduction are numerous. Psychological studies are limited, but several publications have reported on the acceptability, anticipated uptake, positive attitudes, motives, and responses to disclosure of results of genome sequencing. Studies have reported that participants are eager to receive all possible results from their genetic tests (Facio et al., 2013; Hitch et al., 2014; Kaphingst et al., 2016; Meiser, Storey, Quinn, Rahman, & Andrews, 2016; Regier et al., 2015; Sanderson et al., 2016; Wright et al., 2014; Yu, Crouch, Jamal, Bamshad, & Tabor, 2014). Their interest originated from a desire to learn personal health information, including medically actionable results. Also, curiosity and interest in ancestry was often mentioned. In general, participants wished to have a choice about which results from genome sequencing they received (Gollust et al., 2012). Clift et al. (2015) found that cancer patients with failed standard treatment would only receive findings from sequencing that were relevant to their current condition (Clift et al., 2015). Reasons given by patients were that it would cause undue stress and anxiety, that because of religion they would not want to know, and that they were afraid it would cause problems with insurance. Yet another group of patients in their study stated that they did want to obtain all possible findings from exome to find answers and prepare for the future. More recent studies have found that most cancer patients chose to learn results from targeted genome sequencing that could help select a clinical trial, pharmacogenetic and positive prognostic results, and results suggesting inherited predisposition to cancer and treatable noncancer conditions (Gray et al., 2016). The psychological considerations of genome sequencing are discussed, summarized, and challenged throughout this book, alongside thoughtful discussions of the ethical and social responsibilities in using this technology.

CONCLUDING REMARKS

The results of the published studies to date indicate a strong interest in genome sequencing from various patient groups and research participants. Most studies demonstrate that participants most value information about clinically actionable results; patients and participants want to receive information about treatable and/or preventable diseases. There is less consensus about obtaining all information available from genome sequencing. A first impression is that study participants value having a choice in which results they obtain from genome sequencing. Some participants have expressed a sense of inherent ownership about their genomic information, which gives them the right to decide about the results they receive. In the study by Schneider and colleagues (2016), participants stated that a choice should be offered at every level of genomic testing. This includes a choice in the type and quantity of results, but also in how and when to obtain results, and from whom these results can be obtained. In the climate of patient advocacy and the push for patient-reported outcome research, it will be difficult to withhold options from informed clients.

Most people viewed the uncertainty of genome sequencing results as a characteristic of genetic information and were optimistic about sequencing. The primary motives to participate in sequencing studies were to obtain personal health information and to satisfy curiosity. In line with the single-gene testing history, pretest counseling approaches need be developed. Counseling sessions should at least include useful education about genome sequencing. Also, during pretest counseling it should be determined what results patients/participants want to obtain. In the past, the results of single-gene testing were initially treated as potentially dangerous to the wellbeing of patients although this has not been demonstrated by psychological research in the ensuing years. In the era of genome sequencing this assumption should be avoided in the absence of evidence. There already exists evidence from controlled trials that the return of results from genome sequencing that is not threatening can be achieved as effectively as in person from a genetic counselor (Biesecker, Lewis, & Biesecker, 2018). Research will continue to differentiate the need for genetic counseling for medically complex conditions and the return of health-threatening information, from other outcomes of genome sequencing.

There are limitations to the studies that have been conducted to date. Many recruited more highly educated participants and only a couple were representative of the general population (Middleton et al., 2016; Sanderson, Wardle, Jarvis, & Humphries, 2004), limiting generalizability. Initial studies used hypothetical choices to assess intentions to learn results from undergoing genome sequencing. These choices may not reflect the actual decisions people make, as we have seen in the attitudinal studies for Huntington disease before linkage analysis testing began available. Yet the large number of studies that have conducted genome sequencing on participants with study of the outcomes demonstrate minimal negative clinical or psychological outcomes (Robinson et al., under review). Many questions remain to be addressed, but, in the meantime, we should look to what we have learned in genetics to date and give credit to our patients who are able to articulate their needs, desires, and capabilities, and largely make good decisions for themselves.

REFERENCES

Bernhardt, B. A., Roche, M. I., Perry, D. L., Scollon, S. R., Tomlinson, A. N., & Skinner, D. (2015). Experiences with obtaining informed consent for genomic sequencing. *American Journal of Medical Genetics Part A*, *167A*(11), 2635–2646. Available from https://doi.org/10.1002/ajmg.a.37256.

Biesecker, B. B., Klein, W., Lewis, K. L., Fisher, T. C., Wright, M. F., Biesecker, L. G., & Han, P. K. (2014). How do research participants perceive "uncertainty" in genome sequencing? *Genetics in Medicine*, *16*(12), 977–980. Available from https://doi.org/10.1038/gim.2014.57.

Biesecker, B. B., Lewis, K. L., & Biesecker, L. G. (2018). Web-based platform vs genetic counselors in educating patients about carrier results from exome sequencing—Reply. *JAMA Internal Medicine*, *178*(7), 999. Available from https://doi.org/10.1001/jamainternmed.2018.2236.

Biesecker, B. B., & Peters, K. F. (2001). Process studies in genetic counseling: peering into the black box. *American Journal of Medical Genetics*, *106*(3), 191–198. Available from https://doi.org/10.1002/ajmg.10004.

Biesecker, L. G., Burke, W., Kohane, I., Plon, S. E., & Zimmern, R. (2012). Next-generation sequencing in the clinic: Are we ready? *Nature Reviews Genetics*, *13*(11), 818–824. Available from https://doi.org/10.1038/nrg3357.

Bijlsma, R. M., Wessels, H., Wouters, R. H. P., May, A. M., Ausems, M., Voest, E. E., & Bredenoord, A. L. (2018). Cancer patients' intentions towards receiving unsolicited genetic information obtained using next-generation sequencing. *Familial Cancer*, *17*(2), 309–316. Available from https://doi.org/10.1007/s10689-017-0033-7.

Boeldt, D. L., Schork, N. J., Topol, E. J., & Bloss, C. S. (2015). Influence of individual differences in disease perception on consumer response to direct-to-consumer genomic testing. *Clinical Genetics*, *87*(3), 225–232. Available from https://doi.org/10.1111/cge.12419.

Broadstock, M., Michie, S., & Marteau, T. (2000). Psychological consequences of predictive genetic testing: A systematic review. *European Journal of Human Genetics*, *8*(10), 731–738. Available from https://doi.org/10.1038/sj.ejhg.5200532.

Bunnik, E. M., Janssens, A. C., & Schermer, M. H. (2014). Informed consent in direct-to-consumer personal genome testing: The outline of a model between specific and generic consent. *Bioethics*, *28*(7), 343–351. Available from https://doi.org/10.1111/bioe.12004.

Cacioppo, J. T., & Petty, R. E. (1982). The need for cognition. *Journal of Personality and Social Psychology*, *42*, 116–131.

Cacioppo, J. T., Petty, R. E., Feinstein, J. A., & Jarvis, W. B. G. (1996). Dispositional differences in cognitive motivation: The life and times of individuals varying in need for cognition. *Psychological Bulletin*, *119*, 197–253.

Clift, K. E., Halverson, C. M., Fiksdal, A. S., Kumbamu, A., Sharp, R. R., & McCormick, J. B. (2015). Patients' views on incidental findings from clinical exome sequencing. *Applied & Translational Genomics*, *4*, 38–43. Available from https://doi.org/10.1016/j.atg.2015.02.005.

Craufurd, D., MacLeod, R., Frontali, M., Quarrell, O., Bijlsma, E. K., Davis, M., ... Working Group on Genetic Counselling and Testing of the European Huntington's Disease Network (EHDN). (2015). Diagnostic genetic testing for Huntington's disease. *Practical Neurology*, *15*(1), 80–84. Available from https://doi.org/10.1136/practneurol-2013-000790.

Duisterhof, M., Trijsburg, R. W., Niermeijer, M. F., Roos, R. A., & Tibben, A. (2001). Psychological studies in Huntington's disease: Making up the balance. *Journal of Medical Genetics*, *38*(12), 852–861.

Facio, F. M., Brooks, S., Loewenstein, J., Green, S., Biesecker, L. G., & Biesecker, B. B. (2011). Motivators for participation in a whole-genome sequencing study: Implications for translational genomics research. *European Journal of Human Genetics*, *19*(12), 1213–1217. Available from https://doi.org/10.1038/ejhg.2011.123.

Facio, F. M., Eidem, H., Fisher, T., Brooks, S., Linn, A., Kaphingst, K. A., ... Biesecker, B. B. (2013). Intentions to receive individual results from whole-genome sequencing among participants in the ClinSeq study. *European Journal of Human Genetics*, *21*(3), 261–265. Available from https://doi.org/10.1038/ejhg.2012.179.

Fraser, F. C. (1974). Genetic counseling. *The American Journal of Human Genetics*, *26*(5), 636–659.

Goldenberg, A. J., Dodson, D. S., Davis, M. M., & Tarini, B. A. (2014). Parents' interest in whole-genome sequencing of newborns. *Genetics in Medicine, 16*(1), 78−84. Available from https://doi.org/10.1038/gim.2013.76.

Gollust, S. E., Gordon, E. S., Zayac, C., Griffin, G., Christman, M. F., Pyeritz, R. E., ... Bernhardt, B. A. (2012). Motivations and perceptions of early adopters of personalized genomics: Perspectives from research participants. *Public Health Genomics, 15*(1), 22−30. Available from https://doi.org/10.1159/000327296.

Gray, S. W., Park, E. R., Najita, J., Martins, Y., Traeger, L., Bair, E., ... Joffe, S. (2016). Oncologists' and cancer patients' views on whole-exome sequencing and incidental findings: Results from the CanSeq study. *Genetics in Medicine, 18*(10), 1011−1019. Available from https://doi.org/10.1038/gim.2015.207.

Gusella, J. F., Wexler, N. S., Conneally, P. M., Naylor, S. L., Anderson, M. A., Tanzi, R. E., et al. (1983). A polymorphic DNA marker genetically linked to Huntington's disease. *Nature, 306*(5940), 234−238.

Hall, T. O., Renz, A. D., Snapinn, K. W., Bowen, D. J., & Edwards, K. L. (2012). Awareness and uptake of direct-to-consumer genetic testing among cancer cases, their relatives, and controls: The Northwest Cancer Genetics Network. *Genetic Testing and Molecular Biomarkers, 16*(7), 744−748. Available from https://doi.org/10.1089/gtmb.2011.0235.

Hitch, K., Joseph, G., Guiltinan, J., Kianmahd, J., Youngblom, J., & Blanco, A. (2014). Lynch syndrome patients' views of and preferences for return of results following whole exome sequencing. *Journal of Genetic Counseling, 23*(4), 539−551. Available from https://doi.org/10.1007/s10897-014-9687-6.

International Huntington Association and the World Federation of Neurology Research Group on Huntington's Chorea. (1994). Guidelines for the molecular genetics predictive test in Huntington's disease. *Journal of Medical Genetics, 31*(7), 555−559.

Kalia, S. S., Adelman, K., Bale, S. J., Chung, W. K., Eng, C., Evans, J. P., ... Miller, D. T. (2017). Recommendations for reporting of secondary findings in clinical exome and genome sequencing, 2016 update (ACMG SF v2.0): A policy statement of the American College of Medical Genetics and Genomics. *Genetics in Medicine, 19*(2), 249−255. Available from https://doi.org/10.1038/gim.2016.190.

Kaphingst, K. A., Ivanovich, J., Biesecker, B. B., Dresser, R., Seo, J., Dressler, L. G., ... Goodman, M. S. (2016). Preferences for return of incidental findings from genome sequencing among women diagnosed with breast cancer at a young age. *Clinical Genetics, 89*(3), 378−384. Available from https://doi.org/10.1111/cge.12597.

Kessler, S. (1997). Psychological aspects of genetic counseling. IX. Teaching and counseling. *Journal of Genetic Counseling, 6*(3), 287−295. Available from https://doi.org/10.1023/A:1025676205440.

Kessler, S. (2001). Psychological aspects of genetic counseling. XIV. Nondirectiveness and counseling skills. *Genetic Testing, 5*(3), 187−191. Available from https://doi.org/10.1089/10906570152742227.

Kruglanski, A. W. (1989). *Lay epistemics and human knowledge: Cognitive and motivational bases.* New York: Plenum.

MacLeod, R., & Tibben, A. (2014). Genetic counseling and testing. In O. U. Press (Ed.), *Huntington's disease* (4th ed., pp. 165−181). New York: Oxford University Press.

MacLeod, R., Tibben, A., Frontali, M., Evers-Kiebooms, G., Jones, A., Martinez-Descales, A., ... Working Group 'Genetic Testing Counselling' of the European Huntington Disease, N. (2013). Recommendations for the predictive genetic test in Huntington's disease. *Clinical Genetics, 83*(3), 221−231. Available from https://doi.org/10.1111/j.1399-0004.2012.01900.x.

McGuire, A. L., & Beskow, L. M. (2010). Informed consent in genomics and genetic research. *Annual Review of Genomics and Human Genetics, 11*, 361−381. Available from https://doi.org/10.1146/annurev-genom-082509-141711.

Meiser, B., & Dunn, S. (2000). Psychological impact of genetic testing for Huntington's disease: An update of the literature. *Journal of Neurology, Neurosurgery, and Psychiatry, 69*(5), 574−578.

Meiser, B., Storey, B., Quinn, V., Rahman, B., & Andrews, L. (2016). Acceptability of, and information needs regarding, next-generation sequencing in people tested for hereditary cancer: A qualitative study. *Journal of Genetic Counseling, 25*(2), 218−227. Available from https://doi.org/10.1007/s10897-015-9861-5.

Middleton, A., Morley, K. I., Bragin, E., Firth, H. V., Hurles, M. E., Wright, C. F., ... study, D. D. D. (2016). Attitudes of nearly 7000 health professionals, genomic researchers and publics toward the return of incidental results from sequencing research. *European Journal of Human Genetics*, *24*(1), 21–29. Available from https://doi.org/10.1038/ejhg.2015.58.

Nance, M. A. (2017). Genetic counseling and testing for Huntington's disease: A historical review. *American Journal of Medical Genetics B Neuropsychiatr Genet*, *174*(1), 75–92. Available from https://doi.org/10.1002/ajmg.b.32453.

Paulsen, J. S., Nance, M., Kim, J. I., Carlozzi, N. E., Panegyres, P. K., Erwin, C., ... Williams, J. K. (2013). A review of quality of life after predictive testing for and earlier identification of neurodegenerative diseases. *Progress in Neurobiology*, *110*, 2–28. Available from https://doi.org/10.1016/j.pneurobio.2013.08.003.

Quaid, K. A. (2017). Genetic testing for Huntington disease. *Handbook of Clinical Neurology*, *144*, 113–126. Available from https://doi.org/10.1016/b978-0-12-801893-4.00010-9.

Regier, D. A., Peacock, S. J., Pataky, R., van der Hoek, K., Jarvik, G. P., Hoch, J., & Veenstra, D. (2015). Societal preferences for the return of incidental findings from clinical genomic sequencing: A discrete-choice experiment. *Canadian Medical Association Journal*, *187*(6), E190–E197. Available from https://doi.org/10.1503/cmaj.140697.

Resta, R. G. (2006). Defining and redefining the scope and goals of genetic counseling. *American Journal of Medical Genetics Part C: Seminars in Medical Genetics*, *142C*(4), 269–275. Available from https://doi.org/10.1002/ajmg.c.30093.

Robinson, J.O., Wynn, J., Biesecker, B., Bernhardt, B., Brothers, K.B., Chung, W.K., ... Gray, S.W. Psychological Outcomes Related to Exome and Genome Sequencing Result Disclosure: A Meta-Analysis of Seven Clinical Sequencing Exploratory Research (CSER) Consortium Studies. Genetics in Medicine (Under Review).

Sanderson, S. C., Linderman, M. D., Suckiel, S. A., Diaz, G. A., Zinberg, R. E., Ferryman, K., ... Schadt, E. E. (2016). Motivations, concerns and preferences of personal genome sequencing research participants: Baseline findings from the HealthSeq project. *European Journal of Human Genetics*, *24*(1), 153. Available from https://doi.org/10.1038/ejhg.2015.179.

Sanderson, S. C., Wardle, J., Jarvis, M. J., & Humphries, S. E. (2004). Public interest in genetic testing for susceptibility to heart disease and cancer: A population-based survey in the UK. *Preventive Medicine*, *39*(3), 458–464. Available from https://doi.org/10.1016/j.ypmed.2004.04.051.

Sapp, J. C., Dong, D., Stark, C., Ivey, L. E., Hooker, G., Biesecker, L. G., & Biesecker, B. B. (2014). Parental attitudes, values, and beliefs toward the return of results from exome sequencing in children. *Clinical Genetics*, *85*(2), 120–126. Available from https://doi.org/10.1111/cge.12254.

Schneider, J. L., Goddard, K. A., Davis, J., Wilfond, B., Kauffman, T. L., Reiss, J. A., ... McMullen, C. (2016). "Is it worth knowing?" Focus group participants' perceived utility of genomic preconception carrier screening. *Journal of Genetic Counseling*, *25*(1), 135–145. Available from https://doi.org/10.1007/s10897-015-9851-7.

Shahmirzadi, L., Chao, E. C., Palmaer, E., Parra, M. C., Tang, S., & Gonzalez, K. D. (2014). Patient decisions for disclosure of secondary findings among the first 200 individuals undergoing clinical diagnostic exome sequencing. *Genetics in Medicine*, *16*(5), 395–399. Available from https://doi.org/10.1038/gim.2013.153.

Skirton, H., & Patch, C. (2009). *Genetics for the health sciences*. Bloxham: Scion Publishing Ltd.

The Huntington's Disease Collaborative Research Group. (1993). A novel gene containing a trinucleotide repeat that is expanded and unstable on Huntington's disease chromosomes. *Cell*, *72*(6), 971–983.

Tibben, A. (2007). Predictive testing for Huntington's disease. *Brain Research Bulletin*, *72*(2-3), 165–171. Available from https://doi.org/10.1016/j.brainresbull.2006.10.023.

van Dooren, S., Rijnsburger, A. J., Seynaeve, C., Duivenvoorden, H. J., Essink-Bot, M. L., Tilanus-Linthorst, M. M., ... Tibben, A. (2004). Psychological distress in women at increased risk for breast cancer: The role of risk perception. *European Journal of Cancer*, *40*(14), 2056–2063. Available from https://doi.org/10.1016/j.ejca.2004.05.004.

Vos, J., Menko, F. H., Oosterwijk, J. C., van Asperen, C. J., Stiggelbout, A. M., & Tibben, A. (2013). Genetic counseling does not fulfill the counselees' need for certainty in hereditary breast/ovarian cancer families: An explorative assessment. *Psychooncology*, *22*(5), 1167−1176. Available from https://doi.org/10.1002/pon.3125.

Vos, J., Otten, W., van Asperen, C., Jansen, A., Menko, F., & Tibben, A. (2008). The counsellees' view of an unclassified variant in BRCA1/2: Recall, interpretation, and impact on life. *Psychooncology*, *17* (8), 822−830. Available from https://doi.org/10.1002/pon.1311.

Webster, D. M., & Kruglanski, A. W. (1994). Individual differences in need for cognitive closure. *Journal of Personality and Social Psychology*, *67*(6), 1049−1062.

Wright, M. F., Lewis, K. L., Fisher, T. C., Hooker, G. W., Emanuel, T. E., Biesecker, L. G., & Biesecker, B. B. (2014). Preferences for results delivery from exome sequencing/genome sequencing. *Genetics in Medicine*, *16*(6), 442−447. Available from https://doi.org/10.1038/gim.2013.170.

Yu, J. H., Crouch, J., Jamal, S. M., Bamshad, M. J., & Tabor, H. K. (2014). Attitudes of non-African American focus group participants toward return of results from exome and whole genome sequencing. *American Journal of Medical Genetics Part A*, *164A*(9), 2153−2160. Available from https://doi.org/10.1002/ajmg.a.36610.

CHAPTER 2

Genome Sequencing and Individual Responses to Results

Leslie G. Biesecker[1] and Barbara B. Biesecker[2]

[1]Medical Genomics and Metabolic Genetics Branch, National Human Genome Research Institute, National Institutes of Health, Bethesda, MD, United States
[2]Distinguished Fellow, RTI International, Bethesda, MD, United States

Genome sequencing expands the reach of genetic testing by interrogating the entire genome (Biesecker & Green, 2014). This new technology has facilitated the diagnosis of rare and undiagnosed conditions, identified genes that contribute to complex autism, cancer, and cardiovascular disease, identified pharmaceutical effectiveness and adverse reactions, and allowed tumor profiling for treatment selection. It also unveils rare secondary findings for medically actionable results among those undergoing sequencing for another reason (Green et al., 2013; Kalia et al., 2017). Sequencing has been used to identify carrier status and other results deemed valuable by research participants that may have more personal than clinical utility (Kohler, Turbitt, & Biesecker, 2017; Kohler, Turbitt, Lewis, et al., 2017). This chapter reviews genetic sequencing and its capabilities and reports on the peer-reviewed literature that has explored clinical and psychological responses from those undergoing sequencing.

USE OF SEQUENCING AND HOW IT EXTENDS BEYOND GENETIC TESTING

Most genetic tests use sequencing techniques. The term "sequencing" can refer to any of a number of a broadly useful set of techniques that are used in the research and clinical laboratory to assess DNA, the molecular basis of heritable diseases. Because it is broadly useful, it can be used to assess a wide range of genetic testing questions and it is critical to be clear and specific about which particular type of sequencing assay or test is under consideration. Sequencing techniques can be focused or random in nature and nonspecific or directed. In addition to the heterogeneity of sequencing, there is a spectrum of heritable diseases, which can be categorized by the nature of the disease and the type of genetic alteration(s) that are associated with each kind of trait. At one end of the spectrum are heritable diseases that are associated with very rare (sometimes unique) changes in the DNA and if a patient has that single DNA change (variant),

Clinical Genome Sequencing
DOI: https://doi.org/10.1016/B978-0-12-813335-4.00002-7

2019 Published by Elsevier Inc.

there is a high probability that the person will have the trait or disease. Most of the diseases associated with rare variants are also rare. These diseases are inherited in recognizable familial patterns, the so-called Mendelian traits. At the other end of the spectrum are heritable diseases that are associated with common changes in the DNA, variants that may be present in up to 50% of the general population. If a person has one of these DNA variants, it confers a small, but elevated, probability of having the heritable disease associated with that variant. Typically, there are multiple changes in the DNA, scattered throughout the genes that must co-occur, each contributing to risk for the disease, for the disease to manifest. Diseases associated with multiple common variants are themselves common (e.g., type II diabetes mellitus) and are not inherited in recognizable familial patterns, but they do tend to cluster in families. These different kinds of heritable diseases or traits are typically tested in distinct ways, using different technologies.

Until about 2010, rare variants were detecting by use of Sanger sequencing (Smith et al., 1986). This is a directed, focused, low-throughput technique that is nevertheless highly reliable and considered by some to be the "gold standard" for many sequence-based genetic tests. In Sanger sequencing, the protein-coding portions of genes are amplified from a patient DNA sample by a series of polymerase chain reactions (PCR). This is followed by the sequencing reaction where each amplified DNA segment is tagged with four different colored dyes at the end of the strand. The molecules are separated by size and dye color to derive the sequence of the DNA segment, which a software program translates into a colored electropherogram and text file. Because it is a directed, focused technique, the ordering clinician and the laboratory geneticist control what gene is being evaluated.

After 2010, rare variant testing began transitioning to so-called next-generation sequencing. This is fundamentally a nondirected, random, high-throughput technique that takes all of the DNA from a patient sample, shears it randomly into small pieces. These pieces are then attached to a reaction chamber that is basically a double-layer microscope slide, amplified by PCR while attached to the slide, and then each of these millions of pieces of DNA is sequenced simultaneously, one basepair at a time in the slide chamber, using a slightly different set of labeling dyes. This generates hundreds of millions of short (150–200 basepair) sequence reads, which are then analyzed by a computer to reconstruct almost of the DNA sequence of the genome (about 3 billion basepairs) for that individual. This is termed genome sequencing (incorrectly referred to as "whole" genome sequencing). There are a number of variations of this technique whereby it can be more focused, such as exome sequencing (incorrectly referred to as "whole" exome sequencing), which selects for just the protein-coding regions of almost (about 20,000) genes. The critical concept to recognize about these techniques is their comprehensiveness, which is another way of saying unfocused or nondirected. It is both the power and the challenge of these testing techniques that they assess almost genes—for essentially all known heritable diseases. There are even more focused modifications of next-generation sequencing such as gene panels, whereby the protein-coding regions of a set of 10—100 genes are pulled from the patient DNA sample and are thus of intermediate scale—between the Sanger sequencing single-gene test and the comprehensive approach of genome or exome sequencing.

Common variants can be detected by genome sequencing, but not very well by exome sequencing, as they tend to lie outside of the protein-coding portions of genes. However, they can also be detected by chip-based genotyping. Chip testing is based on the premise that because these variants are common, we know what most of them are, so a high-throughput test can be designed to test for these specific variants, which is more economical than sequencing all of the DNA. Economical chip-based tests can be designed to test for millions of such variants on a single chip. Chip-based testing and common disease testing are not further considered in this chapter, but it is essential for the reader to understand the differences between this versus rare-variant, next-generation sequencing.

An important concept is that the above techniques are useful for testing for what are considered small DNA variants—most commonly a single basepair change, but "small" here means up to 50 basepairs in size. Larger variants are called copy number variants because the variation is either a gain or loss of a segment of DNA of 50 basepairs up to hundreds of thousands of basepairs. Copy number variants are assayed using one of two types of chip-based testing—a chip like that described above or what is called a comparative genomic hybridization chip (Bejjani & Shaffer, 2008). Copy number variants can be rare or common and can be associated with a large range of rare and common diseases.

The utility, the novelty, and the psychosocial challenges that arise in newer genetic testing techniques primarily relate to the nondirected, genome-wide techniques of exome and genome sequencing. But they can arise in the other technologies to a lesser degree or frequency. Several of these techniques are used in various combinations in the clinical situations described below.

CLINICAL UTILITY OF RESULTS FROM SEQUENCING
Diagnosis of rare diseases and disease susceptibility

This is a common indication for genome sequencing or exome sequencing commonly accompanied by testing for copy number variants. The benefit of this testing is that its breadth addresses the fundamental diagnostic challenge, which is that rare genetic diseases are difficult to diagnose clinically, which means that the ordering clinician often does not know which single-gene Sanger sequencing test should be ordered. If one has a test that evaluates nearly all genes, one does not need to clinically diagnose the patient to order the test, because the test is the same for essentially all inherited diseases. Autism and intellectual disability are common examples of this because they are nonspecific phenotypes. Autism and intellectual disability can be caused by variants in any one of hundreds of different genes and there are few features of these disorders that would allow the clinician to select the correct single gene, Sanger sequencing test. The diagnostic yield of this testing depends on the specific indication. For severe intellectual disability, multiple congenital anomalies, and several other indications, the diagnostic yield is 20%—30% (Biesecker & Green, 2014), which is remarkably high considering that prior to sequencing, very few could be diagnosed with a specific disorder. The diagnostic yield for other indications can be higher or lower, but irrespective of those specific numbers, the testing allows the interpreting clinician to make a specific, clinical-molecular diagnosis, which is highly useful for determining recurrence risks, predicting the natural history for the affected patient, and in some cases, identifying a treatment based on the molecular diagnosis.

It is important to recognize that many disorders for which exome or genome sequencing is important can also be caused by copy number variants. These are best tested for by one of the methods described above, so many diagnostic approaches include a combination of exome/genome sequencing and a copy number variant chip test, sometimes in a staged approach, sometimes simultaneously. Methods to recognizing copy number variants from genome sequencing data are developing rapidly, but are not yet reliable.

Tumor identification and targeted treatment

Cancer is a genomic disease and typically arises first as an acquired (not inherited) single mutation in a few cells of the body, which perturbs a key cell growth or other regulatory pathway, that is then followed by additional mutations. This mutational progression is correlated with the progression from normal tissue, to benign tumor, to malignant tumor, and finally to metastasis. Importantly, identifying the mutations that are in the cancer can be used to precisely diagnose the cancer and to identify potential treatments specifically targeted to that individual cancer. This is often termed precision cancer therapeutics. To do this, the testing typically includes some combination of exome or genome sequencing, panel testing, or chip genotyping of a noncancer tissue (usually blood, for patients who do not have cancer of blood cells) and the cancer tissue DNA. Then, the two variant datasets are compared (cancer and noncancer) to identify the variants that are in the cancer, but are not in the noncancer sample (so-called tumor-normal sequencing). The variants that are in the tumor but not in the normal sample can be important for therapeutics, but not for understanding the possible inherited basis of a susceptibility. The variants that are in the normal sample, which may or may not be in the tumor, can be useful for understanding the patient's inherited risk for cancer. It is therefore critical to be specific, when discussing cancer sequencing, as to whether one is considering the tumor variants or the nontumor variants with respect to their clinical utility. It is also important to appreciate that many cancer biology laboratories (clinical and research) discard or ignore the variants that are in the normal sample, as they are only interested in the biology and therapeutics of the cancer. The variants from the normal sample have exactly the same utility as do the variants from exome and genome sequencing for other inherited diseases. The variants that are only in the tumor are of much more limited utility and have fewer implications for inherited disease and other consequences of genetic testing.

Carrier testing and screening

Carrier testing is mostly focused on diseases that are inherited in an autosomal recessive pattern, where one must have both of their two copies of a gene harboring disease-causing variants to have the disease. People with only one of two copies of a gene with such a variant are termed carriers, and therefore, in almost all cases, an affected patient only results from a mating of two carriers. Evaluation of carrier status differs from diagnosis of disease or cancer genomics because the tested patient does not have a disease, though they may or may not have a family history of one of these diseases. If they do have a family history, it is termed carrier testing and if they do not, it is termed carrier screening. Another important attribute of carrier testing is that it is of very little clinical utility when performed on an individual—it is primarily useful for pairs of individuals who are considering having children. This is because the risk of having an affected offspring is present only if the couple are both carriers for variants in the same gene—if they are only carriers for a different set of genes, there is essentially no risk for having these diseases. Therefore, carrier results only have utility when one takes into account a couple.

Testing for carriers can be performed using a number of the sequencing or chip-based techniques described above and can be an incidental or secondary use of testing data when a patient is sequenced because they do have a disease. The utility of carrier testing is manifold. It can be used preconceptually by a couple to decide whether or not to have children, it is used in a few religious sects to arrange marriages, it can be used for prenatal diagnosis which sometimes offers the option of selective termination, and it can be used for preimplantation diagnosis with selection of unaffected gametes for implantation in in vitro fertilization.

Secondary findings

Secondary findings, also known as incidental findings, are a set of potential findings from primarily genome and exome sequencing (but uncommonly also from chip genotyping and copy number variant testing) that have a potentially high clinical utility.

The utility is that these are commonly underdiagnosed diseases for which there are medical interventions that are highly effective to reduce morbidity and/or mortality. The basic approach to secondary findings in the United States is that they should be evaluated and, if present, returned to the participant, irrespective of why the sequencing or chip testing was done in the first place (Green et al., 2013; Kalia et al., 2017). On average, about 2% of people who undergo exome or genome sequencing will have a secondary finding variant. Secondary findings are considered a form of screening called opportunistic screening, which is to distinguish it from population screening. Population screening is a directed activity, based on deep, sound epidemiologic and economic analyses from which it is known that there is an overall improvement in population health at an acceptable cost (Wilson & Jungner, 1968). Opportunistic screening is the process of evaluating a result because it was generated for some other reason or no reason at all (people who undergo sequencing for curiosity). It is analogous to chest X-rays and lung cancer. It has been demonstrated that population screening for lung cancer by chest X-rays is not medically or cost-effective. That is, a person without symptoms of lung cancer should not have a chest X-ray for the sole purpose of screening for lung cancer. Yet, every chest X-ray that is performed, regardless of the reason, is evaluated for lung cancer, which is an opportunistic screen for lung cancer.

The acceptance of secondary findings analysis among countries varies widely. In some it is routine, in some it is variable, and in a few it is essentially prohibited. This debate is complex, but suffice it to say that there are valid ethical, financial, and other healthcare system-based arguments for and against secondary findings. For those situations where secondary findings are returned, the issue of which results should be returned is also complex. Most recommendations center on highly medically actionable results, but there are arguments that this should be broadened to include results with potential personal utility, even if there is no clinical utility. It should also be noted that some consider secondary findings to be the deliberate search for such variants, whereas incidental findings are findings in the same genes that are not systematically evaluated, but instead may or may not be inadvertently discovered in a subset of people who harbor them, a random or arbitrary occurrence. While there may be sound reasons to have a policy to disclose or not disclose secondary findings, the arbitrary approach of incidental findings only when stumbled upon should be discouraged because it is inherently capricious and unfair.

Early reports of secondary findings describe high acceptance rates to receive results. While no one prefers to be identified as having a predisposition to sudden cardiac death, such as is the risk for a number of heritable cardiovascular conditions, most participants report feeling grateful for having the opportunity to be closely monitored, eliminate engagement in certain sports activities, and have their children tested (Hart et al., 2018; Sapp et al., 2018). Most participants in genome sequencing studies who are identified to have a pathogenic variant pursue follow-up medical care. Yet those who enroll in research are often more formally educated and motivated to pursue follow-up care than members of the public. There are not yet reports of uptake and utility of secondary findings in clinical sequencing, so the generalizability from the research sequencing cohorts is unknown. The ability to recognize a health risk before a condition alarmingly strikes an unsuspecting individual, is at the heart of the advantages of all presymptomatic diagnostic opportunities, such as the well-accepted practice of widespread newborn screening.

Small studies have identified high levels of communication of secondary finding to relatives (Hart et al., 2018; Sapp et al., 2018), however beyond nuclear families, whether at-risk relatives follow through with genetic testing or are identified to have the familial gene variant has been inconsistent. Communication to relatives and follow-up risk management will be essential to realizing the full benefits of presymptomatic identification of medically actionable gene variants.

Pharmacogenetic testing

Pharmacogenetic traits lie in a gray area between a disease, a susceptibility, and a trait. Many drugs are metabolized by enzymes such that the activity of the drug is reduced (true for many such traits) or enhanced (true for only a few) by this metabolism reaction. Critically, the ability to carry out these kinds of drug metabolic reactions varies among individuals and this variation is heritable and coded for by variants in the genes for these metabolizing enzymes. Most of these traits are not considered diseases because they have no medical effect, unless the person ingests the relevant drug. Also, because it is useful and important for drugs to be metabolized, they are not considered diseases. However, some pharmogenetic traits can make patients very ill or even cause death [e.g., Stevens—Johnson syndrome and carbamazepine, which is an effective seizure drug (Manolio et al., 2018)]. Most pharmacogenetics traits are tested for by chip genotyping, although it is possible to derive such information from genome and exome sequencing (Ng et al., 2017). In that respect, they can also be considered as secondary findings, in that they can be derived from a genome or exome sequence done on an individual for a rare disease.

There are a range of overlapping genetic testing techniques that in one way or another assess sequence variations in the genome. These tests are used medically for a spectrum of overlapping clinical scenarios and they provide a range of overlapping kinds of clinical utility. Thus, when considering the psychosocial responses to sequence-based testing, one must consider the nuances and distinctions of the various forms of testing, the medical testing setting, and the utility of the test and its other implications.

Psychological responses to results

Dr. L. Biesecker leads the US National Institutes of Health's (NIH) ClinSeq cohort studies. The initial cohort of nearly 1000 primarily white and highly educated participants piloted used clinical sequencing at the NIH (Biesecker et al., 2009). Participants completed a baseline survey to assess their understanding and expectations of genome sequencing (Facio et al., 2013). A number of psychological variables on the survey were assessed in relation to participants' intentions to receive sequencing results (Reid et al., 2018; Taber, Klein, Ferrer, Han, et al., 2015; Taber, Klein, Ferrer, Lewis, et al., 2015). Our studies demonstrated that the vast majority of participants intended to receive at least some types of results, and many all results, made available to them. Participants were high in dispositional optimism and reported low levels of ambiguity about learning results from sequencing (Taber, Klein, Ferrer, Han, et al., 2015; Taber, Klein, Ferrer, Lewis, et al., 2015). Participants with greater avoidance of health-threatening information were somewhat less interested in receiving results. And perceptions of social norms predicted intentions to learn results (Reid et al., 2018). This series of studies on intentions to learn results reveals a number of variables theoretically predicted to affect decisions to learn results.

Pathogenic variants in known Mendelian genes, risks for cardiovascular disease, medically actionable secondary findings, and carrier results have all been returned to ClinSeq participants and published (Biesecker et al., 2018; Lewis et al., 2016, 2018). Overall, high intensions to receive results were actualized by high uptake of opportunities to learn results. Understanding of results was high among this well-educated cohort and follow-up health care was pursued when recommended, although communication of results with relatives was not consistently reported (Lewis et al., 2016). The largest set of results returned was carrier results offered via a randomized controlled trial where use of an interactive platform to return results was found to be noninferior to in-person return by a genetic counselor (Biesecker et al., 2018; Lewis et al., 2018). These results were not health-threatening, as participants were beyond their reproductive years, but offered results that were relevant to their adult children. There were no significant differences in psychological outcomes between the two intervention groups and overall anxiety, depressive symptoms, and test-related distress were low. Overall the initial ClinSeq cohort is composed of highly motivated early adopters of new technology who are likely resilient to many potential adverse outcomes of genome sequencing. While the findings are overall reassuring, they are insufficient to generalize to other populations. The second ClinSeq cohort is composed of over 500 black participants of primarily African descent (Lewis et al., 2018). While a more diverse population with more female than male participants, they are relatively highly educated. Investigation of this cohort's response to return of results may lead to a broader range of clinical and psychological outcomes.

While concern persists about adverse psychological outcomes from receipt of genome sequencing results, early evidence from multiple studies demonstrates no significant negative consequences from return of sequence results (Robinson et al., under review). Like the introduction of all new technologies, caution is warranted as evidence continues to gather. Yet in many ways, remarkable adverse psychological consequences of genome sequencing are not expected given that it represents an expansion of genetic testing rather than an altogether new source of information. Genome sequencing does come with challenges that are enumerated throughout this book, yet these pertain primarily to the volume of results and the limitations in interpreting them. From these challenges emerge debate about ways to properly consent individuals (see Chapter 3: Consenting Patients to Genome Sequencing), set reasonable expectations for patients and research participants, and help recipients manage uncertain findings (see Chapter 5: Uncertainties in Genome Sequencing). Across seven NIH-funded US clinical sequencing studies, from the Clinical Sequencing Exploratory Research Consortium, 1300 individuals and parents of affected children completed surveys pre- and post-return of results to capture the psychological impact of receiving a variety of types of findings from genome sequencing (Robinson et al., under review). Robinson and colleagues conducted a meta-analysis across studies to assess change in state anxiety, depressive symptoms, and test-specific results that demonstrated no significant increases in anxiety or depressive symptoms after receipt of results from genome sequencing. Assessment of test-specific distress yielded differences in perceptions of uncertainty and in positive responses to test results, suggesting that measurement at a more granular level is needed to detect subtle differences in psychological outcomes, although these may not rise to the level of clinical significance. Importantly, no adverse negative psychological outcomes were reported. This meta-analysis highlights the importance of arriving at consensus on constructs of greatest relevance and harmonizing use of scales across studies to facilitate comparative analyses and meta-analysis of data.

Outcomes from return of results from smaller genome sequencing studies have also demonstrated relatively positive psychological outcomes (Lewis et al., 2016; Sanderson et al., 2017). This is particularly true in the case of rare undiagnosed disease where results identify a long sought-after etiology (Krabbenborg et al., 2016; Yanes, Humphreys, McInerney-Leo, & Biesecker, 2017). Even in the absence of treatment or intervention, having an explanation for a condition offers patients and parents an avenue to affirm that their condition is "real." Yet having a label without a therapeutic intervention can lead to replacing one uncertainty, the lack of a diagnosis, with another, limited understanding of how the condition may be managed. Uncertainties persist in the expanding identification of conditions, genetic contributions to common disease, and gene—gene and gene—environment interactions. These unknowns will exist for decades to come as we reside in a time of genetic discoveries and determination of their relation to health and development. To help patients and the public acquire realistic expectations of this evolving science, messages to anticipate uncertainties for years to come rather than viewing genetics as a means to certainty are warranted.

Reporting to relatives and cascade testing

When a pathogenic variant is reported to a patient or parent of a child with a rare disorder, one of the subsequent benefits of the information is the opportunity to alert at-risk relatives. Upon learning of their risk, relatives can seek out help from a genetics clinic to undergo single-gene testing (in many cases) to learn whether they too are at increased risk to be affected, develop a condition, and/or to pass on the variant to offspring. Genetic testing of relatives is referred to as cascade testing. One way that the clinical utility of genome sequencing will be evaluated is by whether it leads to cascade testing. Harkening back to experiences with single-gene testing, results have not always been passed along by patients, or parents of patients, to relatives (Forrest et al., 2003). There are many reasons patients may not notify relatives, from being unclear who is at risk, not wanting to alarm their relatives or cause them to feel badly about having or having passed on a familial variant. In some families, relatives do not stay in touch or communicate regularly. Taken together, these reasons often can result in inaction. Communication of genetic risk in families has been studied most often in cancer genetics, where communication of familial risk is not consistently successful. In breast and ovarian cancer families reporting within families has been more successful in hereditary breast and ovarian cancer families than in hereditary colon cancer families, for example. Evidence-based interventions to facilitate family communication of risk will be important to actualize the important outcome of cascade testing from genetic testing and genome sequencing.

It is at times unclear whose responsibility it is to notify relatives and practice varies across countries. In the United States, the responsibility is placed on the patient or parents to notify relatives. Genetic counselors and other healthcare providers make strong efforts to encourage notification and support patients in doing so by supplying draft notification letters that can be personalized and sent to relatives. Such resources help boost patients' confidence in conveying the information accurately and making sound suggestions to relatives for following up on the risk information. Yet even with these efforts, communication can be thwarted. It can be awkward for a patient to become a family informant during a time of seeking treatment for oneself. Generally, communication and cascade testing represent an area where genetic testing has not fully reached its potential to reduce morbidity and mortality. Investigation into interventions to enhance both communication of familial risk and follow-up testing are needed. In other countries, the healthcare system takes on the responsibility to alert at-risk relatives, for example, in the Netherlands, results of abnormal familial hypercholester-olemia genetic tests lead directly to home visits of relatives by visiting nurses to initiate cascade testing (Umans-Eckenhausen, Defesche, Sijbrands, Scheerder, & Kastelein, 2001).

PERSONAL UTILITY

Alongside clinical utility of genomics, a related concept of personal utility has emerged (Kohler, Turbitt, & Biesecker, 2017; Kohler, Turbitt, Lewis, et al., 2017). Even for genetic testing, patient benefits that extend beyond clinical utility have been acknowl-edged (Foster, Mulvihill, & Sharp, 2009; Lupo et al., 2016; Ravitsky & Wilfond, 2006). With the arrival of genome sequencing, a broader conversation about the per-sonal utility of results has emerged. While the definition of personal utility is still being deliberated, it generally describes the perceived benefits of sequencing results beyond those that are clinical and psychological (Kohler, Turbitt, & Biesecker, 2017; Kohler, Turbitt, Lewis, et al., 2017). Early efforts to define this concept have resulted in a number of areas where end-users find benefit in receipt of genome sequencing infor-mation: satisfying curiosity, knowing more about oneself, making informed reproduc-tive decisions, gaining control over one's future, and valuing knowledge. Most of these perceived benefits exist beyond what is typically considered clinical utility and raise questions about the role they play in directing the offer of genome sequencing. On their own, they may not justify access to genome sequencing within a healthcare system, but taken together with potential health benefits, they represent areas that may motivate individuals to pursue sequencing and to engage in follow-up to clinically meaningful results to improve their health. They also demonstrate a broader array of expectations of sequencing that may need to be modified to be realistic.

Defining what constitutes the construct of personal utility, anything beyond clinical utility, is difficult. Theoretically it has endless bounds. It raises the question of whether people should be able to access genome sequencing for nonhealth-related reasons and whether it actually is primarily a medical tool or, more broadly, a source of informa-tion about oneself. This relates to the strong interest in the United States in ancestry testing and identification of biological relatives (see Chapter 6, Direct to Consumer Genome Sequencing, for further discussion of direct-to-consumer testing).

The future use of genome sequencing for health

This chapter reviews the various types of clinical utility that exist and that will further emerge from genome sequencing. Many of the ways it is being used, such as diagnosis of rare diseases, are extensions of the utility of genetic testing, but include ventures into new areas such as pharmacogenetics. Many prestigious scientists project that genome sequencing will become a routine tool in medical care in the near future

(Childs & Valle, 2000; Collins, Green, Guttmacher, & Guyer, 2003). While there are several existing examples of its utility outlined herein, they represent the tip of the iceberg of future genome knowledge. The majority of genes are not understood and those that are understood are often unknown in how they will be expressed or how to offset their negative health consequences. It will be in the coming decades that these unknowns are tackled and medical science advanced. In the meantime, the remainder of this volume addresses the many psychological and ethical considerations of applying genome sequencing to health at this time in our history.

REFERENCES

Bejjani, B. A., & Shaffer, L. G. (2008). Clinical utility of contemporary molecular cytogenetics. *Annual Review of Genomics and Human Genetics, 9*, 71−86.

Biesecker, L. G., & Green, R. C. (2014). Diagnostic clinical genome and exome sequencing. *New England Journal of Medicine, 370*(25), 2418−2425.

Biesecker, L. G., Mullikin, J. C., Facio, F. M., Turner, C., Cherukuri, P. F., Blakesley, R. W., ... Green, E. D. (2009). The ClinSeq project: Piloting large-scale genome sequencing for research in genomic medicine. *Genome Research, 19*(9), 1665−1674.

Biesecker, B. B., Lewis, K. L., Umstead, K. L., Johnston, J. J., Turbitt, E., Fishler, K. P., ... Biesecker, L. G. (2018). Web platform vs in-person genetic counselor for return of carrier results from exome sequencing: A randomized clinical trial. *JAMA Internal Medicine, 178*(3), 338−346.

Childs, B., & Valle, D. (2000). Genetics, biology and disease. *Annual Review of Genomics Human Genetics, 1*(1), 1−19.

Collins, F. S., Green, E. D., Guttmacher, A. E., & Guyer, M. S. (2003). A vision for the future of genomics research. *Nature, 422*(6934), 835−847.

Facio, F. M., Eidem, H., Fisher, T., Brooks, S., Linn, A., Kaphingst, K. A., ... Biesecker, B. B. (2013). Intentions to receive individual results from whole-genome sequencing among participants in the ClinSeq study. *European Journal of Human Genetics, 21*(3), 261−265.

Forrest, K., Simpson, S., Wilson, B., Van Teijlingen, E., McKee, L., Haites, N., & Matthews, E. (2003). To tell or not to tell: Barriers and facilitators in family communication about genetic risk. *Clinical Genetics, 64*(4), 317−326.

Foster, M. W., Mulvihill, J. J., & Sharp, R. R. (2009). Evaluating the utility of personal genomic information. *Genetics in Medicine, 11*(8), 570−574.

Green, R., Berg, J., Grody, W., Kalia, S., Korf, B., Martin, C., ... Ormond, K. (2013). ACMG recommendations for reporting of incidental findings in clinical exome and genome sequencing. *Genetics in Medicine, 15*(7), 565−574.

Hart, M. R., Biesecker, B. B., Blout, C. L., Christensen, K. D., Amendola, L. M., Bergstrom, K. L., ... Hindorff, L. A. (2018). Secondary findings from clinical genomic sequencing: prevalence, patient perspectives, family history assessment, and health-care costs from a multisite study. *Genetics in Medicine*. Available from https://doi.org/10.1038/s41436-018-0308-x, [Epub ahead of print].

Kalia, S. S., Adelman, K., Bale, S. J., Chung, W. K., Eng, C., Evans, J. P., ... Korf, B. R. (2017). Recommendations for reporting of secondary findings in clinical exome and genome sequencing, 2016 update (ACMG SFv2. 0): A policy statement of the American College of Medical Genetics and Genomics. *Genetics in Medicine, 19*(2), 249−255.

Kohler, J. N., Turbitt, E., & Biesecker, B. B. (2017). Personal utility in genomic testing: A systematic literature review. *European Journal of Human Genetics, 25*(6), 662−668.

Kohler, J. N., Turbitt, E., Lewis, K., Wilfond, B. S., Jamal, L., Peay, H. L., ... Biesecker, B. B. (2017). Defining personal utility in genomics: A Delphi study. *Clinical Genetics, 92*(3), 290−297.

Krabbenborg, L., Vissers, L., Schieving, J., Kleefstra, T., Kamsteeg, E., Veltman, J., . . . Van der Burg, S. (2016). Understanding the psychosocial effects of WES test results on parents of children with rare diseases. *Journal of Genetic Counseling*, *25*(6), 1207–1214.

Lewis, K., Hooker, G., Connors, P., Hyams, T., Wright, M., Caldwell, S., . . . Biesecker, B. (2016). Participant use and communication of findings from exome sequencing: A mixed-methods study. *Genetics in Medicine*, *18*(6), 577–583. Available from https://doi.org/10.1038/gim.2015.133.

Lewis, K. L., Umstead, K. L., Johnston, J. J., Miller, I. M., Thompson, L. J., Fishler, K. P., . . . Biesecker, B. B. (2018). Outcomes of counseling after education about carrier results: A randomized controlled trial. *American Journal of Human Genetics*, *102*(4), 540–546.

Lupo, P. J., Robinson, J. O., Diamond, P. M., Jamal, L., Danysh, H. E., Blumenthal-Barby, J., . . . Green, R. C. (2016). Patients' perceived utility of whole-genome sequencing for their healthcare: Findings from the MedSeq project. *Personalized Medicine*, *13*(1), 13–20.

Manolio, T. A., Hutter, C. M., Avigan, M., Cibotti, R., Davis, R. L., Denny, J. C., . . . Shear, N. H. (2018). Research directions in genetic predispositions to Stevens-Johnson syndrome/toxic epidermal necrolysis. *Clinical Pharmacology and Therapeutics*, *103*(3), 390–394. Available from https://doi.org/10.1002/cpt.890.

Ng, D., Hong, C. S., Singh, L. N., Johnston, J. J., Mullikin, J. C., & Biesecker, L. G. (2017). Assessing the capability of massively parallel sequencing for opportunistic pharmacogenetic screening. *Genetics in Medicine*, *19*(3), 357–361. Available from https://doi.org/10.1038/gim.2016.105.

Ravitsky, V., & Wilfond, B. S. (2006). Disclosing individual genetic results to research participants. *The American Journal of Bioethics*, *6*(6), 8–17.

Reid, A. E., Taber, J. M., Ferrer, R. A., Biesecker, B. B., Lewis, K. L., Biesecker, L. G., . . . Klein, W. M. P. (2018). Associations of perceived norms with intentions to learn genomic sequencing results: Roles for attitudes and ambivalence. *Health Psychology*, *37*(6), 553–561.

Robinson, J., Wynn, J., Biesecker, B., Brothers, K., Patrick, D., Rini, C., . . . Gray, S. (under review). A meta-analysis of psychological outcomes across 8 clinical sequencing exploratory research consortium studies.

Sanderson, S. C., Linderman, M. D., Suckiel, S. A., Zinberg, R., Wasserstein, M., Kasarskis, A., . . . Schadt, E. E. (2017). Psychological and behavioural impact of returning personal results from whole-genome sequencing: The HealthSeq project. *European Journal of Human Genetics*, *25*(3), 280–292.

Sapp, J. C., Johnston, J. J., Driscoll, K., Heidlebaugh, A. R., Sagardia, A. M., Dogbe, D. N., . . . NISC Biesecker, L. G. (2018). Evaluation of Recipients of Positive and Negative Secondary Findings Evaluations in a Hybrid CLIA-Research Sequencing Pilot. Comparative Sequencing Program. *American Journal of Human Genetics*, *103*(3), 358–366. Available from https://doi.org/10.1016/j.ajhg.2018.07.018.

Smith, L. M., Sanders, J. Z., Kaiser, R. J., Hughes, P., Dodd, C., Connell, C. R., . . . Hood, L. E. (1986). Fluorescence detection in automated DNA sequence analysis. *Nature*, *321*(6071), 674–679.

Taber, J. M., Klein, W. M., Ferrer, R. A., Han, P. K., Lewis, K. L., Biesecker, L. G., . . . Biesecker, B. B. (2015). Perceived ambiguity as a barrier to intentions to learn genome sequencing results. *Journal of Behavioral Medicine*, *38*, 715–726.

Taber, J. M., Klein, W. M., Ferrer, R. A., Lewis, K. L., Biesecker, L. G., & Biesecker, B. B. (2015). Dispositional optimism and perceived risk interact to predict intentions to learn genome sequencing results. *Health Psychology*, *34*, 718–728.

Umans-Eckenhausen, M. A., Defesche, J. C., Sijbrands, E. J., Scheerder, R. L., & Kastelein, J. J. (2001). Review of first 5 years of screening for familial hypercholesterolaemia in the Netherlands. *Lancet*, *357* (9251), 165–168.

Wilson, J. M. G., & Jungner, G. (1968). *Principles and practice of screening for disease*. Geneva: World Health Organization Retrieved from . Available from http://www.who.int/bulletin/volumes/86/4/07-050112BP.pdf.

Yanes, T., Humphreys, L., McInerney-Leo, A., & Biesecker, B. (2017). Factors associated with parental adaptation to children with an undiagnosed medical condition. *Journal of Genetic Counseling*, *26*(4), 829–840.

FURTHER READING

Lawal, T. A., Lewis, K. L., Johnston, J. J., Heidlebaugh, A. R., Ng, D., Gaston-Johansson, F. G., ... Biesecker, L. G. (2018). Disclosure of cardiac variants of uncertain significance results in an exome cohort. *Clinical Genetics*, *93*(5), 1022–1029.

Taber, J. M., Klein, W. M. P., Lewis, K. L., Johnston, J. J., Biesecker, L. G., & Biesecker, B. B. (2018). Reactions to clinical reinterpretation of a gene variant by participants in a sequencing study. *Genetics in Medicine*, *20*(3), 337–345.

Turbitt, E., Roberts, M. C., Ferrer, R. A., Taber, J. M., Lewis, K. L., Biesecker, L. G., ... Klein, W. M. (2018). Intentions to share exome sequencing results with family members: Exploring spousal beliefs and attitudes. *European Journal of Human Genetics*, *26*(5), 735–739.

CHAPTER 3

Consenting Patients to Genome Sequencing

Emilia Niemiec and Heidi Carmen Howard
Public Health and Caring Sciences, Centre for Research Ethics and Bioethics, Uppsala University, Uppsala, Sweden

INFORMED CONSENT: AN INTRODUCTION

Obtaining informed consent (IC) for a medical intervention or testing is considered as one of the crucial elements of responsible medical and research practice, including genetic and genomic testing. Grady (2015) summarized that: "Informed consent is a process of communication between the health care provider or investigator and the patient or research participant that ultimately culminates in the authorization or refusal of a specific intervention or research study. According to the American Medical Association, 'Informed consent is a basic policy in both ethics and law that physicians must honor . . .'".

The formalized concept of IC was introduced to research, and subsequently to medical practice, to a large extent in reaction to abuses in human subject research in the 20th century, including cruel experiments performed by scientists with eugenic ideals, who attempted to apply and advance knowledge about heredity (Manson & O'Neill, 2007). Over time, the discussions around IC have tended to focus on the respect for autonomy, which was to be secured by IC (Faden & Beauchamp, 1986; Manson & O'Neill, 2007). The literature on consent often outlines the following elements as being necessary for the IC process: decision-making capacity of a patient, information disclosure and its understanding, voluntariness, and authorization (e.g., by signing the form) (Faden & Beauchamp, 1986).

The practice of obtaining consent gradually became implemented in everyday research and clinical practice, as well as entrenched in national legislations (Hoeyer, 2009). Importantly, professional guidelines and law cases have continued to influence the discussion on IC and the practices of obtaining IC, including in genetics and genomics (Borry et al., 2015; Spatz, Krumholz, & Moulton, 2016; Vears, Niemiec, Howard, & Borry, 2018). Over time, the implementation of IC in practice

Clinical Genome Sequencing
DOI: https://doi.org/10.1016/B978-0-12-813335-4.00003-9

has prompted new questions, discussions, and studies. These have concerned, among others, the purpose, meaning, and various functions of IC, conditions for obtaining truly genuine IC, and problems such as those related to placing too much weight on the informational aspect of IC (Manson & O'Neill, 2007; Sugarman et al., 1999). For example, above and beyond the purposes to help support the rights and wellbeing of patients and research participants, IC may also protect healthcare professionals and institutions from liability (Grady, 2015).

Diagnostic analysis of DNA was first introduced to clinical practice in the 1960s, when methods allowing for diagnosis of chromosomal abnormalities were developed (Harper, 2008). Since then, in the context of medical genetics, ethical issues, including IC, seemed to have received particular consideration, which may have been partly caused by a desire to distance current practices of genetics from past misuses of genetic knowledge in the name of eugenic ideology (Nuffield Council on Bioethics, 1993). It may have also been related to the fact that, to some, genetic information is considered to possess distinctive characteristics, thereby making it "exceptional" compared to other medical information (e.g., it is familial and non-changing) (Manson & O'Neill, 2007).[1] Indeed, ethical principles and requirements in genetics, including IC, have been *specifically* addressed in various legal documents, for example, in international legislations (Council of Europe, 2008) and national laws (Kalokairinou et al., 2018). In such a way, the requirement of IC was expanded to genetics, covering not only invasive medical procedures which may result in a physical harm, but also procedures of acquisition and handling of genetic information (Manson & O'Neill, 2007). The challenges related to IC specifically in genetics and genomics and the related issues have been widely discussed in the academic literature and have been addressed by experts and professional societies in guidelines and recommendations (Ayuso, Millán, Mancheño, & Dal-Ré, 2013; Borry et al., 2015).

The challenges discussed in the literature regarding IC and "traditional" genetic testing (GT) (analyzing one or a few genetic variants at a time) include the familial and probabilistic nature of (some) genetic information, the high expectations of patients and the potential problems with understanding the nature of genetic

---

[1] In the beginning of the Human Genome Project, as the result of its associated ELSI (ethical, legal and social implications) program, a "Draft Genetic Privacy Act" was written, including the requirements for informed consent for genetic analysis. Importantly, in the introduction of the document its authors articulated a debatable approach of genetic exceptionalism, stating that "*The Act is based on the premise that genetic information is different from other types of personal information in ways that require special protection*"; in its remainder the document reemphasized uniqueness of genetic information (http://web.ornl.gov/sci/techresources/Human_Genome/resource/privacyact.pdf). As explained by Manson and O'Neill: "Although, the Draft Genetic Privacy Act didn't become part of US legislation, the issues it addressed and the approach it took influenced the subsequent attempts to legislate to protect 'genetic privacy'" (Manson & O'Neill, 2007).

information, as well as gaps in preparation of physicians for offering testing (James, Geller, Bernhardt, Docksum, & Holtzman, 1998; Rimer, Sugarman, Winer, Bluman, & Lerman, 1998). Importantly, genetic counseling has been considered an important element of the IC process for GT in helping to facilitate informed choice (Rieger & Pentz, 1999). The issues surrounding IC in genomics overlap with these and are amplified by the huge increase in data generated (Pinxten & Howard, 2014).

The goal of this chapter is to provide an overview of the current salient issues related to obtaining informed consent in clinical genomics including the challenges, as well as perspectives from the normative (or more idealized) as well as from the more practical contexts.

CHALLENGES OF INFORMED CONSENT IN CLINICAL GENOMICS

Compared to traditional GT, genomic sequencing produces huge volumes of sequencing data concurrently (see Chapter 2: Genome Sequencing and Individual Responses to Results). These data introduce additional features, for example, the possibility of obtaining unsolicited findings, which in turn raise additional or amplified ethical, legal, and social challenges to obtaining IC (Pinxten & Howard, 2014). Clearly, many of the issues related to consent in genetics are similar to those found in genomics (ACMG Board of Directors, 2013). However, these may be seen as amplified in genomics and some additional practical issues arise related to the relative novelty of introducing NGS in the clinic, the increase in data volume, the complexity of interpretation, as well as the overall uncertainty surrounding the use of high-throughput approaches in the clinic (Howard & Iwarsson, 2018).

Indeed, the amount of sequence data generated in whole exome sequencing (WES)/whole genome sequencing (WGS) and the associated "data imperative"[2] raise questions about how genomic data should be handled and which findings should be returned to patients. The categories of possible findings and related issues include, among others: variants relevant to the original clinical question, for which we may have variable levels of information/support regarding whether or not they are meaningful to a phenotype (variants of unknown or variable significance); unsolicited findings, which are findings not related to the original reason for testing but which may still have a medical or social relevance (e.g., a highly penetrant variant related to developing cancer, or nonpaternity); relevance of findings for relatives; and the possibility and/or duty of recontacting patients when new information about

---

[2] By "data imperative" we refer to the situation that appears to be emerging in genomics, where all data generated must be used in some way beyond the original purpose for sequencing; either for clinical purposes (e.g., reanalysis/recontact and/or opportunistic screening) or for research purposes. This is a relatively new situation with respect to the volume of data generated, and stakeholders are still trying to establish proper ELSI frameworks.

variant associations are found which may change the significance of a variant; the length and location of data storage, if it will be stored anonymously, any sharing with third parties, and for which purposes (Borry et al., 2015; Pinxten & Howard, 2014). The more items there are to be consented to, potentially the less clear or obvious it may be to patients what exactly they are consenting to. Each of these issues prompts ongoing debates among different stakeholders, including professional societies. Despite a lack of consensus on many of the ELSI and practicalities related to offering WES/WGS in the clinic, there is a trend towards agreement that many of these aspects should be explained (or at least mentioned along with sources for more information) to patients during the IC process, ideally during pretest counseling (ACMG Board of Directors, 2013; Ayuso et al., 2013). Special care needs to be taken to establish a communication process whereby patients can understand all these aspects well enough to go through informed decision-making.

Given these numerous items that need to be addressed, there are challenges to obtaining IC in the clinical context. These include the resources needed to properly address IC, including time needed in the clinic, education of healthcare practitioners (HCPs), and financial resources to support all these activities (Borry et al., 2018). While the process of IC includes or overlaps with other processes/elements present when offering genomic sequencing, such as genetic counseling (see Chapter 8: Genetic Counseling and Genomic Sequencing, and Chapter 10: Genome Sequencing in Prenatal Testing and Screening; Lessons Learned From Broadening the Scope of Prenatal Genetics From Conventional Karyotyping to Whole Genome Microarray Analysis), decision-making (see Chapter 4: Judgment and Decision-Making in Genome Sequencing, and Chapter 10: Genome Sequencing in Prenatal Testing and Screening; Lessons Learned From Broadening the Scope of Prenatal Genetics From Conventional Karyotyping to Whole Genome Microarray Analysis), and signing an IC form, since the former two issues are addressed in separate chapters, we will not focus on these herein.

WHAT SHOULD INFORMED CONSENT FOR GENOMIC SEQUENCING IDEALLY LOOK LIKE? INSIGHTS FROM NORMATIVE DOCUMENTS

In this section we present different normative documents that offer guidance on what IC in genetics (and by default for now, for *genomics*[3]) should ideally look like. With a focus on Europe, we first briefly address different legally binding documents that address consent. We then present recommendations and guidelines issued from

---

[3] What we mean to highlight with this added clause is that some of these documents address consent in genetics as opposed to specifically consent in *genomics*, and that for now, until revisions are made, we consider these documents to also guide what should be done in genomics.

Table 3.1 Summary of selected normative documents (guidelines and recommendations) addressing consent for high-throughput genomic sequencing in the clinical context

Group (country/region)	Document title (year)	Content/scope
ACMG (USA)	Points to consider for informed consent for genome/exome sequencing (2013)	Consent-specific
ESHG (Europe)	Whole-genome sequencing in health care: recommendations of the European Society of Human Genetics (2013)	WGS in clinic
PHG Foundation (UK)	Realising genomics in clinical practice (2014)	Genomics in the clinic
ACMG (USA)	Points to consider in the clinical application of genomic sequencing (2012)	WGS in the clinic
CCMG (Canada)	The clinical application of genome-wide sequencing for monogenic diseases in Canada: Position Statement of the Canadian College of Medical Geneticists (2015)	WGS in the clinic for monogenic diseases
EuroGenTest Report (Europe)	Guidelines for diagnostic next-generation sequencing (2014)	NGS in the clinic
EuroGenTest Article (Europe)	Guidelines for diagnostic next-generation sequencing (2016)	NGS in the clinic
The Presidential Commission for the Study of Bioethical Issues (USA)	Privacy and progress in whole genome sequencing 2012	WGS and privacy

professional societies and groups (Table 3.1), which specifically address genetics and/or genomics in the clinic.

Legislation addressing informed consent

International guidance

The Convention for the Protection of Human Rights and Dignity of the Human Being with regard to the Application of Biology and Medicine: Convention on Human Rights and Biomedicine also known as the Oviedo Convention (treaty series no. 164, Oviedo, 1997) specifically addresses consent in Chapter II and is legally binding for 29 countries which have signed and ratified it (https://www.coe.int/en/web/conventions/full-list/-/conventions/treaty/164/signatures?p_auth = HzxJmZTX). In Article 5, the General Rules for interventions in the health field include: (1) the need for free and IC; (2) providing "appropriate information as to the purpose and nature of the intervention as well as on its consequences and risks"; and that (3) consent can be freely withdrawn at any time. Additional specifications are provided in the remaining articles of that chapter for persons unable to consent, the protection of persons who have mental disorders, emergency situations, and previously expressed wishes (Council of Europe, 1997).

Importantly, Article 12 in Chapter IV of the convention addresses specifically predictive GT. It states, among others, that such testing should be the "subject of appropriate genetic counselling" (Council of Europe, 1997).

The Additional Protocol to the Convention on Human Rights and Biomedicine, concerning Genetic Testing for Health Purposes provides further details in Chapter IV Articles 8 and 9 (Council of Europe, 2008). While genetic counseling is explicitly identified in Article 8, implicitly much of this is directed to the obtaining of IC as well:

"Information and genetic counseling:

1. When a genetic test is envisaged, the person concerned shall be provided with prior appropriate information in particular on the purpose and the nature of the test, as well as the implications of its results.

2. For predictive genetic tests as referred to in Article 12 of the Convention on Human Rights and Biomedicine, appropriate genetic counselling shall also be available for the person concerned. The tests concerned are:
   — tests predictive of a monogenic disease,
   — tests serving to detect a genetic predisposition or genetic susceptibility to a disease,
   — tests serving to identify the subject as a healthy carrier of a gene responsible for a disease.

The form and extent of this genetic counselling shall be defined according to the implications of the results of the test and their significance for the person or the members of his or her family, including possible implications concerning procreation choices. Genetic counselling shall be given in a non-directive manner" (Council of Europe, 2008).

Clearly, both the Oviedo Convention and the Additional Protocol place an importance not only on adequate information being provided before a genetic test, but also on the manner in which the information should be offered (e.g., during counseling and in a nondirective manner).

Article 9, then goes on to explicitly identify the need to obtain IC:

"Article 9—Consent

1. A genetic test may only be carried out after the person concerned has given free and informed consent to it. Consent to tests referred to in Article 8, paragraph 2,[4] shall be documented.

2. The person concerned may freely withdraw consent at any time" (Council of Europe, 2008).

---

[4] See list above.

These two articles together clearly state the foundational importance of the consent process as a tool to help support and protect the rights of patients who receive GT.

European Union regulations

Much of the offer of GT in the clinic may be controlled as a healthcare service. As such, the types of tests offered, the facilities where the testing can take place, and the qualifications of the person offering the test may all be restricted. Often, these areas are addressed via national legislation (see the section below: National legislations in European countries) (Kalokairinou et al., 2018). That being said, GT for medical purposes can also be considered as a product and as such can be addressed, among others, via laws on medical devices under the competence of the EU. In particular, in Europe, Directive 98/79 EC on In Vitro Diagnostic (IVD) Medical Devices has the aim of ensuring safety and efficiency of IVD devices offered in the European market (European Parliament & the Council of the European Union, 1998; Niemiec, Kalokairinou & Howard, 2017). The IVD Directive will be replaced by a new IVD Regulation in 2022, in which consent is specified in Article 4 on "Genetic information, counseling and informed consent," whereby it is stated:

> "Member States shall ensure that where a genetic test is used on individuals, in the context of healthcare as defined in point (a) of Article 3 of Directive 2011/24/EU of the European Parliament and of the Council and for the medical purposes of diagnostics, improvement of treatment, predictive or prenatal testing, the individual being tested or, where applicable, his or her legally designated representative is provided with relevant information on the nature, the significance and the implications of the genetic test, as appropriate."
>
> **European Parliament & the Council of the European Union (2017).**

With respect to the offer of GT for health purposes outside the traditional genetic clinic [i.e., direct-to-consumer (DTC)] Kalokairinou also suggests that the harmonization of a regulatory approach around IC (genetic counseling and medical supervision) may help to allow for more effective oversight of DTC GT and in doing so, could help reduce some of the risks associated with DTC GT (Kalokairinou et al., 2018). It is not obvious, however, whether any such harmonized approach would be particularly beneficial for genetic or genomic testing in the clinic.

Finally, it is important to note that personal data are protected in the EU by the General Data Protection Regulation (GDPR), which was adopted by the European Parliament on May 24, 2016, and has been applicable in all EU Members States since May 25, 2018. Much of the (academic) literature to date regarding the GDPR and genomics has focused on trying to explain the impact of the GDRP on the processing of genomic data for research. However, the tight link between clinical and research genomics contexts (see the explanation of "data imperative" in the section: Challenges of Informed Consent in Clinical Genomics), among others, makes it important to carefully consider how the GDPR may affect the processing of genomic data (and consent for this) specifically in the context of the clinic. While comprehensive analysis of this issue is beyond the scope of this chapter, we suggest that readers keep this in mind in the future.

National legislations in European countries

As mentioned above, regulation of clinical practice resides with the Member States rather than the EU; as a result, aspects related to this, such as genetic counseling, etc., are regulated on the national level. Kalokairinou and co-authors recently conducted a pan-European study of how national legislations in 26 European countries specifically deal with, among others, IC for GT. They found that different countries have different levels of specification regarding what information should be given to patients and whether written IC is required (Kalokairinou et al., 2018). The authors conclude that "imposing strict uniform standards of informed consent, genetic counselling and medical supervision by means of an EU Regulation, seems impractical and restrictive and beyond the legislative competence of the EU. While it is important to ensure equal level of protection of patient rights across Europe, including availability of genetic counselling and medical supervision, as well as adequate informed consent processes, such standards can be achieved by Member States following the Additional Protocol on genetic testing for health purposes (Council of Europe, 2008) and relevant international and European guidelines (European Society of Human Genetics 2010; Organisation for Economic Co-operation and Development 2007)" (Kalokairinou et al., 2018).

Professional guidelines and recommendations

A number of professional genetic (and genomic) societies, as well as advisory bodies, have issued guidelines and recommendations on the use of high-throughput genomic approaches in the clinical context (ACMG Board of Directors, 2012; Boycott et al., 2015; Matthijs et al., 2016; van El et al., 2013). These guidelines address a wide scope of issues and report on several aspects, including but not limited to the issues of IC, return of results, and/or the duty to recontact (Table 3.1). Very few such normative documents have specifically focused only on IC in genomic sequencing. Below we firstly address the documents with larger scope followed by those that specifically address consent.

Need to establish a responsible consent process

Many normative documents that address genomic sequencing in general, explicitly state that IC must be adapted and refined to establish an adequate process to deal with the use of NGS in the clinic. There is also a specific call for the sharing of experiences and policies to best establish such a process (van El et al., 2013). This being said, EuroGentest wisely points out that if NGS is just being used to replace Sanger sequencing, and that the results are likely to be the same as using the latter sequencing approach then there is less of a need to adapt consent (as may be the case if only one or a few genes are sequenced, e.g., *BRCA1* and *BRCA2*) (EuroGentest, 2014).

Overlapping areas: genetic counseling and informed consent

It can be difficult to separate obtaining IC from genetic counseling (see Chapter 10: Genome Sequencing in Prenatal Testing and Screening; Lessons Learned From Broadening the Scope of Prenatal Genetics From Conventional Karyotyping to Whole Genome Microarray Analysis). This is the case since the former is usually addressed as part of the latter in the traditional clinical genetic context. In fact, all these normative documents address consent within or related to the process of genetic counseling. Hence, it will be interesting to see if genetic counseling is to change due to the demands of genomics (e.g., if genomic counseling will be recognized as a separate training/service/profession) and if or how this may affect the obtaining of IC. We discuss some empirical studies providing insight into this issue in the section on empirical studies: Stakeholders views and experiences of informed consent process.

Need for explicit informed consent for genomic sequencing and research

Many recommendations underline the need for clear differentiation between consent for clinical use and consent for research (ACMG Board of Directors, 2013; Boycott et al., 2015; Matthijs et al., 2016). Some also suggest the use of a different consent form. EuroGentest suggests that a second counseling session should take place, including the consent process, when going from having analyzed a panel of genes to analyzing the entire genome or exome (EuroGentest, 2014).

Responsibility of different professionals in the process of informed consent: much seems to rest on the genetic clinician yet still a lot of responsibilities remain ill-defined

According to EuroGentest it is crucial that "the referring physician is fully informed about the limitations and possible unfortunate effects of a genetic test" (Matthijs et al., 2016). The laboratory should provide relevant information about the NGS test, however, as EuroGentest explicitly states, it is not the responsibility of the laboratories to develop, provide, or collect IC. Moreover, of particular interest with respect to the ongoing debates over broad and specific consent, this description appears to be more narrow than broad. Meanwhile, in their report on "Realising Genomics in Clinical Practice" the PHG Foundation states in Recommendation 7 that "It is the responsibility of the referring clinician to provide transparent information and to seek consent relating to targeted and open sequencing and analysis" (PHG Foundation, 2014).

The area of responsibility for obtaining IC is a good example of where there is a lack of clarity on the substeps needed to reach the stage of obtaining responsible IC from a patient. For example, who is responsible for helping clinicians/clinical centers develop a proper consent procedure and/or consent forms? Could template forms ever capture enough of the diversity in contexts to be valid and useful for the many different healthcare systems and their idiosyncratic ways of offering WGS/WES? Should consent procedures and/or forms be "approved" to ensure they include all the necessary information and that they are written in a language that can be understood by larger publics? If so, who would conduct the controls/provide approval? Based on what criteria?

What needs to be included in the informed consent process?

Indeed, most normative documents covering the wider scope of genomic sequencing (above and beyond IC), also provide, in a general manner, the criteria for IC.

While already performed some time ago, Ayuso et al. (2013) conducted a review of the literature and offer one of the most exhaustive lists of elements to be included in a consent form or procedure for whole genome sequencing in the clinic. They recommend that the following should be included: (1) pretest counseling by an expert; (2) definition or delimitation of scope of testing; (3) description of testing [procedure, and testing purpose (e.g., screening, presymptomatic, etc.)]; (4) benefits; (5) risks; (6) the voluntary nature of testing; (7) the possibility of refusal at any time; (8) description of alternative diagnostic tests; (9) information pertinent to relatives (and informing them or not); (10) the inclusion of patient data in a database; (11) how privacy and confidentiality will be respected; (12) storage and future (secondary) use of results; (13) how incidental findings will be managed; and (14) the need for specific IC (Ayuso et al., 2013).

Furthermore, some articles by groups of individual authors also offer guidance on IC. Such articles based on empirical approaches are discussed in the following section.

EMPIRICAL STUDIES ON THE PROCESS OF INFORMED CONSENT FOR GENOMIC SEQUENCING

The process of consenting patients to whole genome and exome sequencing has been investigated in empirical studies. While they are not representative of all clinical/research genetics contexts, they provide a much-needed glimpse of what is happening in practice currently and, importantly, how practice may differ from the ideal notions of IC. Notably, most of the studies discussed focus on a particular context, that is of English-speaking countries, which may not be generalizable to other settings; more diverse contexts need to be explored further.

Informed consent documents: what can we learn about practice?

While the IC process is much more than the IC form, the study of the latter can be an important indication of what information is valued by clinicians and/or sequencing providers in the IC process. They also offer a concrete object of analysis when trying to reconcile the ideals of IC with what is happening in practice.

Variety in elements addressed in informed consent forms

A few studies have specifically analyzed the content of IC forms for whole exome or genome sequencing in the clinic and in the context of translational studies (Table 3.2). These analyses report that there exists an important variation in content and presentation of elements of IC addressed in the forms studied.

A study of 29 IC forms by Fowler, Saunders, and Hoffman (2017) shows that some elements of IC tend to be present more often than others. Specifically, they found that all the forms described in some way the uncertainty of results obtained in the sequencing, related to technical limitations of the sequencing or challenges in the interpretation of results (Fowler et al., 2017). Jamal et al. (2013) explained that the variety of approaches in IC forms may be related to the initial (and novel) stage of implementation of genomic sequencing in the clinic (Jamal et al., 2013). Furthermore, with respect to laboratories, Fowler et al. (2017) suggested that they may not yet be aware of the relevant recommendations or may have chosen not to comply with them (Fowler et al., 2017). Notwithstanding, as Jamal et al. (2013) indicated, "this variation has practical implications for healthcare providers and institutions evaluating, comparing, and ultimately deciding which ES (exon sequencing) provider to use" (Jamal et al., 2013). Additionally, current practices regarding IC may serve as templates or examples, which may influence practices of other providers which will start offering genomic sequencing in the near future (Jamal et al., 2013). Therefore, as suggested in the professional recommendations, evidence-based standards for best practices should be developed and implemented (Henderson et al., 2014; Jamal et al., 2013; van El et al., 2013).

Henderson et al. (2014), based on their analysis, provided "a checklist to help identify gaps and resolve ambiguities in consent forms for sequencing." Meanwhile, Fowler et al. (2017) specifically encouraged implementation of the American College of Medical Genetics (ACMG) and the Presidential Commission for the Study of Bioethical Issues guidelines, on which they based their work, to facilitate the communication in the process of IC. The authors also suggested that their work can serve as a checklist facilitating compliance with these guidelines.

Secondary and unsolicited findings in informed consent forms

Both Jamal et al. (2013) and Fowler et al. (2017) reported that the studied documents did not completely address the issues related to secondary findings. Issues of unsolicited

Table 3.2 Summary of publications presenting empirical data on informed consent for high-throughput genomic sequencing

Authors, year	What/who was subject of the study	Type of test for which IC was obtained	Source/context of IC	Approach of the study
Jamal et al. (2013)	Six IC forms (and other documents available such as requisition forms)	WES	US-based CLIA-certified labs	"Core elements" of IC were defined by the authors with involvement of stakeholders; presence of these elements in the IC forms was evaluated. Readability analysis included
Henderson et al. (2014)	Nine IC forms	WES and WGS	CSER funded by NIH	Content analysis using codes based on the initial review of the forms. Readability analysis included
Fowler et al. (2017)	29 IC forms and addenda	WES	US-based CLIA-certified labs	Content analysis using codes based on recommendations of ACMG and PCSBI. Readability analysis included
Niemiec et al. (2016)	Documents to which consumers had to agree in order to undertake the testing	WES and WGS	Four direct-to-consumer genetic testing companies	Content analysis using codes based on elements of IC recommended by Ayuso et al.
Niemiec and Howard (2016)	Documents to which consumers had to agree in order to undertake the testing	WES and WGS	Four direct-to-consumer genetic testing companies	Content analysis using inductive codes relevant to data and samples' usage
Niemiec et al. (2017)	36 IC forms in English retrieved in internet search	WES and WGS	Clinical sequencing offered mostly by English-speaking institutions/laboratories	Mainly readability was studied
Vears et al. (2018)	54 IC forms in English retrieved in internet search	WES, WGS, and large NGS panels	Clinical sequencing offered mostly by English-speaking institutions/laboratories	Content analysis using inductive codes related to secondary and unsolicited findings
Bernhardt et al. (2015)	29 genetic counselors and research coordinators	WES and WGS	CSER and the clinical context, USA	Interviews exploring experiences of obtaining IC. Both inductive and deductive codes: initial codes related to questions asked, new codes were added during the analysis

Study	Participants		Setting	Method
Tomlinson et al. (2016)	29 genetic counselors and research coordinators	WES and WGS	CSER and the clinical context, USA	Interviews exploring most challenging or memorable case of IC. Inductive codes derived from the analysis of the descriptions
Rigter et al. (2014)	12 professional experts, three patients, and three genetic counselors	WES	Diagnostic WES at VU University Medical Center Amsterdam, the Netherlands	Semistructured interviews exploring views on or experiences with IC; observation of IC sessions
Kaphingst et al. (2012)	311 research participants	WGS	ClinSeq—a clinical study at the NIH, USA	Surveys before and after IC measuring knowledge of genomic sequencing. Surveys included statements to which participants responded using Likert scale
Turbitt et al. (2018)	188 research participants, comparison between lower literacy consent forms and standard consent forms	WGS	Research program investigating primary ovarian insufficiency funded by the NIH, USA	Surveys before and after IC measuring knowledge of genomic sequencing, perceived benefits, decisional conflict, and informed choice to learn secondary variants
Benjamin et al. (2016)	Project participants, patients, and public involvement groups, health professional recruiters	WGS	100,000 Genome Project, UK—a research project embedded in the clinical setting	Combination of approaches was used, both qualitative and quantitative
Tolusso et al. (2017)	53 parents who consented to WES for their children	WES	Clinical WES in children ages 0–18 years at the Cincinnati Children's Hospital Medical Center, USA	Questionnaire evaluating actual and perceived understanding of WES

CLIA, Clinical Laboratory Improvement Amendments; CSER, Clinical Sequencing Exploratory Research; DTC GT, direct-to-consumer genetic testing; NIH, National Institutes of Health; PCSBI, Presidential Commission for the Study of Bioethical Issues.

and secondary findings in the context of IC were specifically analyzed in the study of Vears et al. (2018). The analysis of 54 IC forms revealed, among others, inconsistency (and potential confusion) in the use of terms to denote secondary and unsolicited findings. Furthermore, the forms studied varied in the way they addressed reporting these findings, including instances of forms that did not mention these items at all, opposing relevant recommendations (Vears et al., 2018).

Some of these results may reflect the presence of conflicting positions and recommendations regarding unsolicited and secondary findings. Furthermore, the inconsistency among the forms in terminology may be partly due to the changes in applying terms in the recommendations of the ACMG from "incidental findings" to "secondary findings" to denote the results not related to the initial diagnostic question but deliberately searched for during the sequence analysis (ACMG Board of Directors, 2014; Green et al., 2013). Importantly, the information related to return of results, including a question about which results will be returned was indicated by genetic counselors as one of the important elements of IC, which should be understood by patients (Bernhardt et al., 2015; see also Chapter 8: Genetic Counseling and Genomic Sequencing).

Readability of informed consent forms

The other common finding among the studies of IC forms is low readability of the forms, signaling possible difficulties for patients in understanding the content of the forms (Henderson et al., 2014; Jamal et al., 2013; Niemiec, Vears, Borry & Howard, 2017). Readability tests (e.g., Flesch—Kincaid, SMOG) analyze elements, such as sentences and words length, to estimate reading grade levels, that is, number of years of education needed to understand a given text.

Four studies evaluated the readability of IC forms for genomic sequencing (Table 3.2) identifying relatively high reading grade levels of consent forms (expressed in number of years of education needed to understand a given text). Specifically, they reported the following values of reading grade levels: median 10.8 (Henderson et al., 2014), mean of 10.8 (Fowler et al., 2017) (both calculated in the Flesch—Kincaid test), and 40 in Flesh Reading Ease, which corresponds to between high-school and college grade levels (Jamal et al., 2013). Niemiec et al. (2017) analyzed 36 forms focusing specifically on their readability; the median reading grade levels obtained were 12.2 (in the Flesch—Kincaid formula) and 14.75 (in the SMOG formula). All the average and median reading grade level values reported by the studies were above the grade level of 8th, which is recommended for IC (e.g., by Institutional Review Boards of US medical schools) (Fowler et al., 2017; Henderson et al., 2014; Jamal et al., 2013; Niemiec, 2017; Paasche-Orlow, Taylor, & Brancati, 2003). These results indicate that even if IC forms contain the required elements of information, they may not be understandable for many patients, and may even prevent patients from (fully) reading the forms.

Indeed, experiences of genetic counselors and research coordinators confirm that most patients are not able to understand the content of IC forms for genomic sequencing (Bernhardt et al., 2015). While the relevant information may be explained to patients by a HCP during the IC process, given that genomic sequencing may be increasingly offered by primary physicians not trained specifically in genetics, it is still important that both patients and HCPs have access to adequate written information as a reference. The studies of readability underscore also the problems in communicating genomic information, such as those related to complexity of vocabulary used and questions about its adequacy for patients. Initial efforts have been made to explore which kind of vocabulary may be suitable for patients (Parry & Middleton, 2017).

Informed consent for WES/WGS offered directly-to-consumers
Currently, WES/WGS is available also through DTC offerings, that is, via companies which sell testing to consumers without any involvement of a HCP (e.g., Dante labs, https://www.dantelabs.com/), or which target advertisements to consumers, but require the involvement of a HCP in the provision of the testing (Niemiec, Borry, Pinxten, & Howard, 2016). Given that HCPs may increasingly encounter patients' requests for ordering such testing and that some of these tests are advertised as clinical sequencing and/or they have the Clinical Laboratory Improvement Amendments (CLIA) certification, we will also briefly discuss IC in this context. Two studies focus specifically on IC for WGS/WES offered by DTC companies (Niemiec & Howard, 2016; Niemiec et al., 2016). Specifically, information concerning the following elements of IC was studied: pretest counseling, benefits and risks, incidental findings, and storage and use of consumers' samples and data. The revealed concerns include, firstly, the lack of engagement of healthcare professionals in offering of the tests, including lack of pretest counseling. Secondly, some of the companies did not provide adequate IC documents for genomic testing and for research activities on consumers' samples and data. From the standpoint of medical and research ethics, replacing the process of IC with a "clickwrap" agreement seems to be unacceptable (Niemiec & Howard, 2016; Niemiec et al., 2016). Thirdly, the studies revealed that the content of the documents studied raised concerns. These related mainly to the lack of relevant information and/or the presence of potentially misleading descriptions of some aspects of the testing as well as the secondary use of consumer samples and health-related data. Consequently, consumers might not be aware of all the implications of undertaking WGS/WES, including the potential benefits and risks, or the usage of their samples and/or health-related data for research purposes. Therefore, consumers' acceptance or the given consent might not be truly informed (Niemiec & Howard, 2016; Niemiec et al., 2016). Additionally, the lack of transparency in provision of information about the usage of consumers' data for

research could undermine trust in research practices in general, including publicly funded research (Howard, Sterckx, Cockbain, Cambon-Thomsen, & Borry, 2015).

Stakeholders' views and experiences of the informed consent process

Obtaining the views and experiences of different stakeholders regarding consent in genomics may be valuable, among others, in revealing how well the consent process fulfills different stakeholders' needs, and how important or desired the consent process is for different stakeholders.

Healthcare professional views and experience

The analyses of Bernhardt et al. (2015) and Tomlinson et al. (2016) addressed the experiences of 29 genetic counselors and research coordinators with the IC process for genomic sequencing in the clinical context and in the context of a translational study (Clinical Sequencing Exploratory Research funded by the National Institutes of Health) in the United States. The interviewees reported that the IC process involved one or two (in case insurance preauthorization was required) sessions, which lasted between 10 and 70 minutes, with 30 minutes as the most common length (Bernhardt et al., 2015). Families' engagement in the sessions and lack of previous knowledge or exposure to issues related to GT were factors increasing the length of IC sessions. If the IC was obtained in the context of translational research, the sessions were longer, as more elements related to research had to be explained. The interviewees also mentioned that they provided "more global counselling before testing, and more in-depth counselling after testing, based on test results" (Bernhardt et al., 2015).

Furthermore, the genetic counselors and research coordinators interviewed reported flexibility in ways they addressed "standard" IC elements. Over time, with gained experience with obtaining IC, the interviewees tended to adjust the IC sessions based on contextual factors, such as patients' understanding of genetics, previous contacts with GT, indications for sequencing, the stage of their disease, and whether it was a child or an adult sequenced. With time, they conducted the sessions "in a less structured and more conversational manner, a style that they believed promoted better understanding and engagement" (Bernhardt et al., 2015). These findings suggest that HCPs play a crucial role in the IC process, ensuring the adequate and tailored provision of information relevant to IC. This seems to be in line with an analysis of IC forms, which showed, among others, that IC forms often included statements formulating the responsibility of a HCP in explaining relevant information to patients (Niemiec, Vears, Borry & Howard, 2017). Meanwhile, IC forms, as reported by the genetic counselors and research coordinators, seemed to have limited informative function: "interviewees recognized that most patients and participants cannot attend to, let alone understand, all of the information contained in the consent documents" (Bernhardt et al., 2015).

Importantly, the genetic counselors and research coordinators outlined also common misconceptions and concerns of patients and the elements of IC which they believed were most important for patients to understand; these include, among others, limitations of testing and implications of results (Bernhardt et al., 2015). Furthermore, the same group of interviewees was asked to describe particularly challenging case of obtaining IC (Tomlinson et al., 2016). The challenges in these cases were related to either patients' understanding or facilitation of decision-making. The authors suggested that approaches improving understanding and encouraging deliberation such as simplified IC and extended IC discussions should be evaluated in the context of genomic sequencing (Tomlinson et al., 2016). Furthermore, Bernhardt et al. (2015) indicated that processes of developing IC could benefit from involving various stakeholders to define the minimal set of information to be included in the IC process. Tomlinson et al. (2016) also suggested that consensus is needed on the aspects a patient should understand, and the extent to which these should be understood. This is echoed in another study investigating the experiences of professionals and patients regarding IC in the European context (Rigter et al., 2014). The respondents indicated that it is important to assess and document experiences regarding IC for WES in this early stage of its implementation; based on such information, these IC procedures may be modified in the future. Also echoing other studies, the interviewees indicated that the information relevant to IC was complex to convey to patients (Rigter et al., 2014).

Patients' perspective

There are also studies focusing specifically on patients' and research participants' experiences with IC for genomic sequencing. A study of Kaphingst et al. (2012) investigated the knowledge of genome sequencing of 311 participants in a clinical genome sequencing study (ClinSeq) before and after obtaining IC. The results of the study indicated that overall there was an increase in the knowledge about genome sequencing benefits and limitations after the IC process (Kaphingst et al., 2012). Furthermore, the study indicated that some less-educated groups and some racial/ethnic minority groups were less familiar with limitations of genome sequencing prior to the IC process, highlighting their educational/pretest counseling needs.

Recently, Turbitt et al. (2018) conducted a randomized controlled study to compare the effects of using consent forms with purposefully lower literacy (as compared to standard consent forms) for genome sequencing in research. The authors developed the novel IC form of reduced complexity based on the standard NIH consent forms and the Presidential Commission for the Study of Bioethical Issues guidelines (Presidential Commission for the Study of Bioethical Issues, 2012; Turbitt et al., 2018). The new form was piloted among participants who provided their feedback based on which the wording of the document was further simplified. This lower literacy consent form met the requirement of reading grade level 8 or below (scoring 7.9 in the

Flesch–Kincaid test) suggested for IC forms (Paasche-Orlow et al., 2003; Turbitt et al., 2018). The consent sessions were conducted over the phone and followed by a survey 24 hours and 6 weeks after the sessions to measure the knowledge of sequencing benefits, expected personal benefits, and decisional conflict outcomes. No significant differences were found between the groups of participants consented using standard and lower literacy consent forms in these three categories measured at 6 weeks. Yet the usage of the novel form prompted more questions from the participants. The authors concluded that using such a lower literacy form may be beneficial as it encourages interaction between a participant and a healthcare professional, which may facilitate shared decision-making (Turbitt et al., 2018).

IC was also investigated in the context of WES for pediatric patients, who seem to constitute a large group of patients undergoing WES (Yang et al., 2013). A study by Tolusso et al. (2017) focused on parents who consented to WES for their children. The authors investigated perceived and actual understanding of WES of these adults. The results highlighted issues of which parents had low understanding, which included how genes are analyzed and protection against insurance discrimination as well as issues related to secondary findings. The authors suggested that these topics may need additional attention or emphasis, for example, in a form of supplemental material, such as a patient brochure about insurance issues. Furthermore, addressing the probability of a diagnosis in IC sessions may be beneficial so that the parents do not have inflated expectations for the results of WES (Tolusso et al., 2017).

The IC process has also been evaluated in the context of the 100,000 Genome Project—a research project embedded in the clinical setting, having aims of establishing genomic medicine within the National Health Service in the United Kingdom. An evaluation process involving both qualitative and quantitative approaches engaged three groups of stakeholders, that is, project participants, patient and public involvement groups associated with genomic medicine centers, and health professional recruiters. Based on this investigation, recommendations were formulated to, among others, improve the readability and presentation of IC forms, offer an audio/video of the consent for individuals who have difficulties with attending to the written documents, and provide ongoing consent training to the project recruiters (Benjamin et al., 2016).

SPECIFIC CONTEXTS NECESSITATING ADDITIONAL REFLECTION AND STUDY REGARDING INFORMED CONSENT

As this chapter on IC for genomic sequencing in the clinical context aims at providing an overview, it is beyond its scope to attend to specific contexts. However, we provide below three specific contexts, which, by virtue of the vulnerability of the persons involved and/or the novelty or sheer detail of the situations necessitate further (ethical, legal, and social) considerations regarding IC.

The case of minors

The above discussion about obtaining IC for genomic sequencing in the clinic did not address the issue of age or ability to provide IC. Indeed, apart from the many debates about GT in minors (Borry, Goffin, Nys, & Dierickx, 2008) the challenges of obtaining such consent may also be considered to be amplified. The European Society of Human Genetics recommends that for a minor the primary aim of GT should be the child's best interest (Borry, Evers-Kiebooms, Cornel, Clarke, & Dierickx, 2009a; Borry, Evers-Kiebooms, Cornel, Clarke, & Dierickx, 2009b). The recommendations continue by stating that

> *"(2) The opinion of the minor shall be taken into consideration as an increasingly determining factor in proportion to his or her age and degree of maturity. Decision-making involving the health care of a minor should include, to the greatest extent feasible, his or her consent or assent" (Borry et al., 2009b).*

While these recommendations are laudable, they do not provide additional guidance on how to deal with the realities of obtaining consent or assent from the proband. Indeed, since we have yet to fully solve the issue of IC for genomics in adults, it may be seen as "running before walking" to already be offering it in minors. While we do not mean that all ELSI must be fully solved before offering genomic testing to children; we do suggest that, perhaps, to ensure that IC is truly valid, we need to spend more time addressing this process in the case of minors. Furthermore, what of consent for sequencing in newborns? While the notion of using this in a screening program is not upheld by all groups (Howard, Knoppers, et al., 2015), pilot programs in the United States have tried out WES in healthy newborns as well as in severely ill newborns (Holm et al., 2018). In the case of routine newborn screening (i.e., for all newborns, not based on the appearance of symptoms), the IC process would have to change quite drastically from the current situation where the public health benefits are deemed to outweigh the need to obtain IC for newborn screening (Howard, Knoppers, et al., 2015). For severely ill newborns, the consent obtained from parents for WES/WGS is as (un)questionable as it would be for any life-saving procedure; is there any real choice for parents (see also Chapter 9: Genome Sequencing in Pediatrics: Ethical Issues)? Furthermore, the information obtained at that point may change over time, and may be considered to infringe on one's (the child's) right not to know.

Genomic sequencing within the context of gene editing; Russian doll model of complications?

Whenever gene editing is used to alter the sequence of the genome (via deletion or insertion), both in somatic and germline gene editing, whole genome sequencing is subsequently performed on the altered cells to verify for off-target events, which may have taken place at sites other than the desired target site. Such off-target events may cause changes in the genome that result in negative health consequences (e.g., cancer). Importantly, this means that every patient for whom somatic gene editing is conducted [and many hundreds of patients are currently part of such clinical trials (Maeder & Gersbach, 2016)] will also have their genome sequenced. Like the issue of "running before walking" discussed above, here the fact that we will be using two very novel technologies, and the fact that we have not yet fully solved the issues of appropriate consent for WGS amplify the ethical concerns for consent for gene editing. Above and beyond the issues discussed above regarding WGS/WES, issues surrounding enticement (e.g., financial compensation to donate) and risks (e.g., physical burden on women) to donate gametes for germline gene-editing studies, both of which are relevant to the IC, should be further examined and care should be taken to protect patients and research participants.

CONCLUSION

We have presented different perspectives around pertinent issues of obtaining IC for genomic sequencing in the clinical context. We have discussed both the ideals of what IC should be as well as glimpses into what is actually happening in practice. The former was accomplished by presenting the theory behind IC as well as by presenting normative documents on IC in genetics and genomics. The latter was presented by describing some of the main findings from empirical studies on IC in high-throughput GT. While these studies tended to be more exploratory and descriptive in nature and did not attempt to necessarily "grade" the IC process against the recommendations, they do present the challenges (and sometimes solutions) facing stakeholders, and one could conclude that there are, indeed, several gaps between the ideals (e.g., of professional recommendations) and the applied obtaining of IC at the moment. The solutions and guidelines for practitioners indicated on the basis of the empirical studies include, among others:

- To pay attention that consent forms are readable to patients (Niemiec, Vears, Borry & Howard, 2017; Turbitt et al., 2018);
- To use available and relevant checklists of elements to be included in IC in designing IC forms and when obtaining IC (Ayuso et al., 2013; Fowler et al., 2017; Henderson et al., 2014);
- To focus on elements of information relevant to genomic testing which are particularly challenging to patients. The article of Bernhardt et al. (2015) provides a list of them, which is based on the experiences of genetic counselors.

Reasons why there are gaps between theory and practice can be related in part to a general problem of guidelines whereby they indicate an ideal state without detailing the practical steps to reach this ideal (e.g. patients should fully understand the risks and benefits of testing). Another reason for these gaps may be attributed to the fact that despite great strides in the inclusion of genomic sequencing in the clinic (in some countries), this approach is still only beginning to be used more frequently.

With this in mind, it is important to remember that most of the studies described are not quantitative studies and are not meant to provide representative samples with generalizable results. Furthermore, it is crucial to understand that even such quantitative studies in one country or region (with a specific healthcare system) would not (necessarily) represent the approaches in other countries or regions. While the United Kingdom, the Netherlands, and the United States may have incorporated genomic sequencing more regularly in the clinic, this is not the case for many other countries in the American continents as well as Europe. More empirical studies on IC, exploring various contexts and diverse stakeholder groups, especially patients, may facilitate further understanding of the process and ways to improve it. As mentioned above, the top-down process of defining what elements of IC should look like, may not necessarily meet the actual needs and be tailored to the practice. For this reason, engaging practicing professionals and patients in research exploring issues related to the IC process, as well as in the process of designing IC forms and defining its necessary elements, may be invaluable. Research is needed also to explore and understand the functions IC may have for various stakeholder groups, for example, how IC benefits/impacts different actors, for instance, by protecting against liability or empowering in some ways.

Furthermore, it is important to note that given the current "data imperative," which pushes stakeholders to use as much as possible or "maximize" "the potential of" collected sequencing data, clinical genome sequencing will often (if not always) be tightly linked with the use of those data for research. It will be important to ensure that patients not only understand the difference between the clinical and research use of data, but that they are offered meaningful choices with respect to refusing secondary use of their data.

Finally, we would like to emphasize that the approach to IC, its form, and content are dependent on norms and values which guide the practice of genetics and genomics. As described by Manson and O'Neill (2007):

"Consent, we have argued, can be used to waive important norms, rules and standards, and so has considerable ethical importance. But since its use always presupposes whichever norms are to be waived, it cannot be basic to ethics, or to bioethics".

Indeed, it is also important to reflect on and discuss what are and should be the values in genetics and genomics, including questions, such as what constitutes risks and benefits to patients. Research in this area and informed activities aiming at upholding high ethical standards should be ongoing efforts to ensure that new genomic technologies are used for the benefit of humans and any potential harms, especially to vulnerable populations (e.g., minors), should be addressed and minimized or avoided.

ACKNOWLEDGMENTS

Part of this work has been supported by the SIENNA project (Stakeholder-Informed Ethics for New technologies with high socio-ecoNomic and human rights impAct), which has received funding under the European Union's H2020 research and innovation program under grant agreement No. 741716. This chapter and its contents reflect only the views of the authors and do not intend to reflect those of the European Commission. The European Commission is not responsible for any use that may be made of the information this chapter contains.

REFERENCES

ACMG Board of Directors. (2012). Points to consider in the clinical application of genomic sequencing. *Genetics in Medicine, 14*(8), 759−761. Available from https://doi.org/10.1030/gim.2012.74.

ACMG Board of Directors. (2013). Points to consider for informed consent for genome/exome sequencing. *Genetics in Medicine, 15*(9), 748−749. Available from https://doi.org/10.1038/gim.2013.94.

ACMG Board of Directors. (2014). ACMG policy statement: Updated recommendations regarding analysis and reporting of secondary findings in clinical genome-scale sequencing. *Genetics in Medicine, 17*, 68−69. Available from https://doi.org/10.1038/gim.2014.151.

Ayuso, C., Millán, J. M., Mancheño, M., & Dal-Ré, R. (2013). Informed consent for whole-genome sequencing studies in the clinical setting. Proposed recommendations on essential content and process. *European Journal of Human Genetics: EJHG, 21*(10), 1054−1059. Available from https://doi.org/10.1038/ejhg.2012.297.

Benjamin, C., Boudioni, M., Ward, H., Marston, E., Lindenmeyer, A., Bangee, M., ... Dinh, L. (2016). *National service evaluation of the consent process and participant materials used in the 100,000 Genomes Project.* Retrieved from https://www.genomicsengland.co.uk/consent-evaluation/.

Bernhardt, B. A., Roche, M. I., Perry, D. L., Scollon, S. R., Tomlinson, A. N., & Skinner, D. (2015). Experiences with obtaining informed consent for genomic sequencing. *American Journal of Medical Genetics, 167A*(11), 2635−2646. Available from https://doi.org/10.1002/ajmg.a.37256.

Borry, P., Bentzen, H. B., Budin-Ljøsne, I., Cornel, M. C., Howard, H. C., Feeney, O., ... Felzmann, H. (2018). The challenges of the expanded availability of genomic information: An agenda-setting paper. *Journal of Community Genetics, 9*(2), 103−116. Available from https://doi.org/10.1007/s12687-017-0331-7.

Borry, P., Chokoshvili, D., Niemiec, E., Kalokairinou, L., Vears, D., & Howard, H. C. (2015). *Current ethical issues related to the implementation of whole-exome and whole-genome sequencing. Movement Disorder Genetics* (pp. 481−497). Springer International Publishing.

Borry, P., Evers-Kiebooms, G., Cornel, M. C., Clarke, A., & Dierickx, K. (2009a). Genetic testing in asymptomatic minors: Background considerations towards ESHG Recommendations. *European Journal of Human Genetics, 17*(6), 711−719. Available from https://doi.org/10.1038/ejhg.2009.25.

Borry, P., Evers-Kiebooms, G., Cornel, M. C., Clarke, A., & Dierickx, K. (2009b). Genetic testing in asymptomatic minors: Recommendations of the European Society of Human Genetics. *European Journal of Human Genetics, 17*(6), 711−719. Available from https://doi.org/10.1038/ejhg.2009.25.

Borry, P., Goffin, T., Nys, H., & Dierickx, K. (2008). Predictive genetic testing in minors for adult-onset genetic diseases. *The Mount Sinai Journal of Medicine, 75*, 287−296.

Boycott, K., Hartley, T., Adam, S., Bernier, F., Chong, K., Fernandez, B. A., ... Armour, C. M. (2015). The clinical application of genome-wide sequencing for monogenic diseases in Canada: Position statement of the Canadian College of medical geneticists. *Journal of Medical Genetics, 52*(7), 431−437. Available from https://doi.org/10.1136/jmedgenet-2015-103144.

Council of Europe. (1997). *Convention for the protection of human rights and dignity of the human being with regard to the application of biology and medicine: Convention on human rights and biomedicine.* Retrieved from http://conventions.coe.int/Treaty/en/Treaties/Html/164.htm.

Council of Europe. (2008). *Additional protocol to the convention on human rights and biomedicine concerning genetic testing for health purposes.* Retrieved from http://conventions.coe.int/Treaty/en/Treaties/Html/203.htm.

EuroGentest. (2014). *Guidelines for diagnostic next generation sequencing.*

European Parliament and the Council of the European Union. (1998). *Directive 98/79/EC of the European Parliament and of the Council of 27 October 1998 on in vitro diagnostic medical devices (1998).* Retrieved from http://eur-lex.europa.eu/legal-content/EN/TXT/?uri = celex:31998L0079.

European Parliament and the Council of the European Union. (2016). *Regulation (EU) 2016/679 of the European Parliament and of the Council of 27 April 2016 on the protection of natural persons with regard to the processing of personal data and on the free movement of such data, and repealing Directive 95/46/EC (General Data Protection Regulation).* OJ L 119, 4.5.2016 (p. 1–88). Retrieved from http://data.europa.eu/eli/reg/2016/679/oj.

European Parliament and the Council of the European Union. (2017). *Regulation (EU) 2017/746 of the European Parliament and of the Council of 5 April 2017 on in vitro diagnostic medical devices and repealing Directive 98/79/EC and Commission Decision 2010/227/EU.* OJ L 117, 5.5.2017 (pp. 176–332). Retrieved from http://eur-lex.europa.eu/legal-content/EN/TXT/PDF/?uri = CONSIL:PE_15_2017_INIT&rid = 4

Faden, R. R., & Beauchamp, T. L. (1986). *A history and theory of informed consent.* New York: Oxford University Press.

Fowler, S. A., Saunders, C. J., & Hoffman, M. A. (2017). Variation among consent forms for clinical whole exome sequencing. *Journal of Genetic Counseling, 27*(1), 104–114. Available from https://doi.org/10.1007/s10897-017-0127-2.

Grady, C. (2015). Enduring and emerging challenges of informed consent. *New England Journal of Medicine, 372*(9), 855–862. Available from https://doi.org/10.1056/NEJMra1411250.

Green, R. C., Berg, J. S., Grody, W. W., Kalia, S. S., Korf, B. R., Martin, C. L., ... Biesecker, L. G. (2013). ACMG recommendations for reporting of incidental findings in clinical exome and genome sequencing. *Genetics in Medicine, 15*(7), 565–574. Available from https://doi.org/10.1038/gim.2013.73.

Harper, P. S. (2008). *A Short History of Medical Genetics.* Oxford: Oxford University Press, 557 pp.

Henderson, G., Wolf, S., Kuczynski, K., Joffe, S., Sharp, R., Parsons, D., ... Appelbaum, P. S. (2014). The challenge of informed consent and return of results in translational genomics: Empirical analysis and recommendations. *The Journal of Law, Medicine & Ethics, 42*(3), 344–355. Retrieved from http://onlinelibrary.wiley.com/doi/10.1111/jlme.12151/full.

Hoeyer, K. (2009). Informed consent: The making of a ubiquitous rule in medical practice. *Organization, 16*(2), 267–288. Available from https://doi.org/10.1177/1350508408100478.

Holm, I. A., Agrawal, P. B., Ceyhan-Birsoy, O., Christensen, K. D., Fayer, S., Frankel, L. A., ... Beggs, A. H. (2018). The BabySeq project: Implementing genomic sequencing in newborns. *BMC Pediatrics, 18*(1), 1–10. Available from https://doi.org/10.1186/s12887-018-1200-1.

Howard, H. C., & Iwarsson, E. (2018). Mapping uncertainty in genomics. *Journal of Risk Research, 21*(2), 117–128. Available from https://doi.org/10.1080/13669877.2016.1215344.

Howard, H. C., Knoppers, B. M., Cornel, M. C., Wright Clayton, E., Sénécal, K., & Borry, P. (2015). Whole-genome sequencing in newborn screening? A statement on the continued importance of targeted approaches in newborn screening programmes. *European Journal of Human Genetics: EJHG, August 2014,* 1–8. Available from https://doi.org/10.1038/ejhg.2014.289.

Howard, H. C., Sterckx, S., Cockbain, J., Cambon-Thomsen, A., & Borry, P. (2015). The convergence of direct-to-consumer genetic testing companies and biobanking activities. In M. Wienroth, & E. Rodrigues (Eds.), *Knowing new biotechnologies: Social aspects of technological convergence* (pp. 59–74). New York: Routledge.

Jamal, S. M., Yu, J., Chong, J. X., Dent, K. M., Conta, J. H., Tabor, H. K., & Bamshad, M. J. (2013). Practices and policies of clinical exome sequencing providers: Analysis and implications. *American Journal of Medical Genetics, 161*(5), 935–950. Available from https://doi.org/10.1002/ajmg.a.35942.

James, C., Geller, G., Bernhardt, B. A., Docksum, T., & Holtzman, N. A. (1998). Are practicing and future physicians prepared to obtain informed consent? The case of genetic testing for susceptibility to breast cancer. *Community Genetics*, *1*(4), 203−212. Retrieved from http://www.scopus.com/inward/record.url?eid = 2-s2.0-0032227245&partnerID = tZOtx3y1.

Kalokairinou, L., Howard, H. C., Slokenberga, S., Fisher, E., Flatscher-Thöni, M., Hartlev, M., ... Borry, P. (2018). Legislation of direct-to-consumer genetic testing in Europe: A fragmented regulatory landscape. *Journal of Community Genetics*, *9*(2), 117−132. Available from https://doi.org/10.1007/s12687-017-0344-2.

Kaphingst, K. A., Facio, F. M., Cheng, M.-R., Brooks, S., Eidem, H., Linn, A., ... Biesecker, L. G. (2012). Effects of informed consent for individual genome sequencing on relevant knowledge. *Clinical Genetics*, *82*(5), 408−415. Available from https://doi.org/10.1111/j.1399-0004.2012.01909.x.

Maeder, M. L., & Gersbach, C. A. (2016). Genome-editing technologies for gene and cell therapy. *Molecular Therapy*, *24*(3), 430−446. Available from https://doi.org/10.1038/mt.2016.10.

Manson, N. C., & O'Neill, O. (2007). *Rethinking informed consent in bioethics*. Cambridge: Cambridge University Press.

Matthijs, G., Souche, E., Alders, M., Corveleyn, A., Eck, S., Feenstra, I., ... Bauer, P. (2016). Guidelines for diagnostic next-generation sequencing. *European Journal of Human Genetics*. Available from https://doi.org/10.1038/ejhg.2015.226.

Niemiec, E., Kalokairinou, L., & Howard, H. C. (2017). Current ethical and legal issues in health-related direct-to-consumer genetic testing. *Personalized Medicine*, *14*(5), 433−445. Available from https://doi.org/10.2217/pme-2017-0029.

Niemiec, E., Vears, D., Borry, P., & Howard, H. C. (2017). Readability of informed consent forms for whole-exome and whole-genome sequencing. *Journal of Community Genetics*, *9*(2), 143−151. Available from https://doi.org/10.1007/s12687-017-0324-6.

Niemiec, E., Borry, P., Pinxten, W., & Howard, H. C. (2016). Content analysis of informed consent for whole genome sequencing offered by direct-to-consumer genetic testing companies. *Human Mutation*, *37*(12), 1248−1256. Available from https://doi.org/10.1002/humu.23122.

Niemiec, E., & Howard, H. C. (2016). Ethical issues in consumer genome sequencing: Use of consumers' samples and data. *Applied and Translational Genomics*, *8*, 23−30. Available from https://doi.org/10.1016/j.atg.2016.01.005.

Nuffield Council on Bioethics. (1993). *Genetic screening—Ethical issues*. London. Retrieved from http://nuffieldbioethics.org/wp-content/uploads/2014/07/Genetic_screening_report.pdf.

Paasche-Orlow, M. K., Taylor, H. A., & Brancati, F. L. (2003). Readability standards for informed-consent forms as compared with actual readability. *New England Journal of Medicine*, *348*(8), 721−726. Available from https://doi.org/10.1056/NEJMsa021212.

Parry, V., & Middleton, A. (2017). Socialising the genome. *The Lancet*, *389*(10079), 1603−1604. Available from https://doi.org/10.1016/S0140-6736(17)31011-5.

PHG Foundation. (2014). *Realising genomics in clinical practice*. Retrieved from http://www.phgfoundation.org/reports/16447/

Pinxten, W., & Howard, H. C. (2014). Ethical issues raised by whole genome sequencing. Best practice & research. *Clinical Gastroenterology*, *28*(2), 269−279. Available from https://doi.org/10.1016/j.bpg.2014.02.004.

Presidential Commission for the Study of Bioethical Issues. (2012). *Privacy and progress in whole genome sequencing*. Washington, DC. https://doi.org/10.1016/0167-4048(86)90058-1

Rieger, P. T., & Pentz, R. D. (1999). Genetic testing and informed consent. *Seminars in Oncology Nursing*, *15*(2), 104−115.

Rigter, T., van Aart, C. J. A., Elting, M. W., Waisfisz, Q., Cornel, M. C., & Henneman, L. (2014). Informed consent for exome sequencing in diagnostics: Exploring first experiences and views of professionals and patients. *Clinical Genetics*, *85*(5), 417−422. Available from https://doi.org/10.1111/cge.12299.

Rimer, B. K., Sugarman, J., Winer, E., Bluman, L. G., & Lerman, C. (1998). Informed consent for BRCA1 and BRCA2 testing. *Breast Disease*, *10*(1−2), 99−114. Retrieved from http://www.scopus.com/inward/record.url?eid = 2-s2.0-0031922167&partnerID = tZOtx3y1.

Spatz, E., Krumholz, H., & Moulton, B. (2016). The new era of informed consent: Getting to a reasonable-patient standard through shared decision making. *JAMA: Journal of the American Medical Association, 315*(19), 6—7. Available from https://doi.org/10.1001/jama.2016.3070.

Sugarman, J., McCrory, D. C., Powell, D., Krasny, A., Adams, B., Ball, E., & Cassell, C. (1999). Empirical research on informed consent. An annotated bibliography. *The Hastings Center Report, 29*(1), S1—S42. Retrieved from http://search.ebscohost.com/login.aspx?direct = true&db = c8h&AN = 1999033544&site = ehost-live.

Tolusso, L. K., Collins, K., Zhang, X., Holle, J. R., Valencia, C. A., & Myers, M. F. (2017). Pediatric whole exome sequencing: An assessment of parents' perceived and actual understanding. *Journal of Genetic Counseling, 26*, 792—805. Available from https://doi.org/10.1007/s10897-016-0052-9.

Tomlinson, A. N., Skinner, D., Perry, D. L., Scollon, S. R., Roche, M. I., & Bernhardt, B. A. (2016). "Not tied up neatly with a bow": Professionals' challenging cases in informed consent for genomic sequencing. *Journal of Genetic Counseling, 14*(11), 871—882. Available from https://doi.org/10.1007/s10897-015-9842-8.

Turbitt, E., Chrysostomou, P. P., Peay, H. L., Heidlebaugh, A. R., Nelson, L. M., & Biesecker, B. B. (2018). A randomized controlled study of a consent intervention for participating in an NIH genome sequencing study. *European Journal of Human Genetics, 26*, 622—630. Available from https://doi.org/10.1038/s41431-018-0105-7.

van El, C. G., Cornel, M. C., Borry, P., Hastings, R. J., Fellmann, F., Hodgson, S. V., . . . de Wert, G. M. W. R. (2013). Whole-genome sequencing in health care: Recommendations of the European Society of Human Genetics. *European Journal of Human Genetics: EJHG, 21*(6), 580—584. Available from https://doi.org/10.1038/ejhg.2013.46.

Vears, D., Niemiec, E., Howard, H. C., & Borry, P. (2018). How do consent forms for diagnostic high-throughput sequencing address unsolicited and secondary findings? A content analysis. *Clinical Genetics, 94*(3-4), 1—9. Available from https://doi.org/10.1111/cge.13391.

Yang, Y., Muzny, D. M., Reid, J. G., Bainbridge, M. N., Willis, A., Ward, P. A., . . . Eng, C. M. (2013). Clinical whole-exome sequencing for the diagnosis of Mendelian disorders. *The New England Journal of Medicine, 369*(16), 1502—1511. Available from https://doi.org/10.1056/NEJMoa1306555.

FURTHER READING

National Commission for the Protection of Human Subjects of Biomedical and Behavioral Research. (1979). *Belmont report*. Retrieved from https://www.hhs.gov/ohrp/regulations-and-policy/belmont-report/read-the-belmont-report/index.html.

Nuremberg Code. (1949). *Nuremberg code*. Retrieved June 16, 2017, from https://history.nih.gov/research/downloads/nuremberg.pdf.

World Medical Association. (2013). *WMA Declaration of Helsinki—Ethical principles for medical research involving human subjects*. Retrieved August 9, 2018, from https://www.wma.net/policies-post/wma-declaration-of-helsinki-ethical-principles-for-medical-research-involving-human-subjects/.

CHAPTER 4

Judgment and Decision Making in Genome Sequencing

William M.P. Klein[1,3], Barbara B. Biesecker[2] and Erin Turbitt[3]
[1]Division of Cancer Control and Population Sciences, National Cancer Institute, Bethesda, MD, United States
[2]Distinguished Fellow, RTI International, Bethesda, MD, United States
[3]National Human Genome Research Institute, Bethesda, MD, United States

INTRODUCTION

Genetic testing increasingly includes the use of genome sequencing to assess health risks, most commonly in the context of determining heritable cancer and cardiovascular disease risk (Biesecker & Green, 2014). Such testing may be medically recommended, as in the example of undergoing germline panel testing to inform treatment decisions following a diagnosis of breast cancer. A woman without detected heritable risk may be advised to pursue more conservative risk management such as lumpectomy, whereas a woman identified with an increased heritable cancer risk may be counseled to undergo mastectomy and chemoprevention. In contrast, many other testing options in clinical genomics are preference-based, meaning patients decide according to their values, beliefs, and prior experiences.

One of the critical reasons decisions are often preference-based in genomics is the significant uncertainty about how mutations, referred to as pathogenic variants, contribute to disease risk and diagnoses. In many ways, gene panel testing resembles the decades-long history of genetic testing (see Chapter 1: Genetic Testing Expanded), although with a broadening of the information, its uncertainties, and the high likelihood of receiving variants of uncertain significance (VUS). Importantly, genetic testing began as a tool to diagnose rare disease. As technology advanced, testing evolved to apply gene panel testing where multiple genes known to contribute to heritable risk for cancer and cardiovascular disease are tested simultaneously. Genome or exome sequencing is used to analyze the genes on the panels. This application of the technology illustrates the rapid progression from genetic to genomic testing with the added complexities of making decisions to undergo testing and use the information gained.

Genetics providers often engage in shared decision-making to deliberate the relative merits and costs of learning information from genome sequencing. To make an informed choice, the patient makes a decision with sufficient relevant knowledge

Clinical Genome Sequencing
DOI: https://doi.org/10.1016/B978-0-12-813335-4.00004-0

while adhering to related values and beliefs (Marteau, Dormandy, & Michie, 2001). Yet the process of decision-making is complex, involving a number of cognitive, affective, and motivational factors, leading to disparate consequences. For example, receipt of one or more VUS can evoke a range of responses ranging from increasing perceived risk to providing reassurance (Skinner et al., 2018; Werner-Lin et al., 2018). Yet the clinical significance of these results is unknown, lending no evidence toward or away from risk.

Although there is no reason to presume that patients cannot make informed decisions about testing and use the information in productive ways to manage their health risks, there are nevertheless numerous opportunities for distortions in risk perceptions, errors in application of information, and confusion about unexpected information. These challenges in translating genomics into both clinical and research settings introduce opportunities for social and behavioral research that can inform genetic counseling, health services, and related health policies. Research in these areas can suggest applicable theoretical frameworks (Gooding, Organista, Burack, & Biesecker, 2006) and key constructs of relevance.

LESSONS FROM RESEARCH ON JUDGMENT AND DECISION-MAKING

The genomics context is awash with consequential judgments and decisions. Examples of *judgments* are: perceptions of the value of genome sequencing results, assessments of one's risk of contracting a genetic disease, estimates of population penetrance, and forecasts of one's potential reactions to positive test results. *Decisions* are often extrapolations or consequences of judgments and might include whether or not to seek results, what types of results to seek (and when), if and when to disclose results to family, and whether to engage in behavior change to offset any negative effects of a genetic predisposition. A rich literature originating in the fields of decision science, medical decision-making, social psychology, cognitive psychology, and behavioral economics has explored many of the psychological and other behavioral processes that influence how people arrive at judgments and how they make consequential choices, providing a compelling foundation for understanding how people might engage with genome sequencing (see, e.g., Khan et al., 2015). Because only a small fraction of this work has been conducted specifically in the genomics context, we consider relevant literature in multiple contexts while drawing attention to genomics examples where available.

Fundamental to the study of human judgment and decision-making is the supposition—and observation—that people must conserve limited cognitive resources in order to achieve everyday goals and in so doing may rely on time-saving heuristics or "rules of thumb" when making judgments and choices (Tversky & Kahneman, 1974). Use of such heuristics is not necessarily problematic, but can lead to a number of errors and biases due to the use of incomplete or unknown information. Moreover, people's cognitive resources constitute only one set of factors that may influence judgment and decision-making. Emotional states—both those related to a decision (e.g., worry about the results of a genetic test) and those unrelated (e.g., anger due to an unexpected medical charge)—can greatly influence how people think and reason in ways that can either impede or facilitate productive decision-making (e.g., Slovic, Finucane, Peters, & MacGregor, 2007). People also harbor a wide variety of motives that can be said to "hover" over their thought processes—such as the motive to appear rational to oneself and others (Shafir & LeBoeuf, 2002) and to regard oneself positively (Steele, 1988). Importantly, one must consider both the effects of and the interactions among cognitive, emotional, and motivational factors when attempting to understand people's engagement with decision-making in genome sequencing. We outline each in turn below with reference to how these factors have been or could be shown to influence judgment and decision-making in the genome sequencing context.

Cognitive biases

Research on the cognitive underpinnings of judgment and decision making has highlighted several consistent and fundamental conclusions about human reasoning that are relevant to many decision-making contexts. In particular, people seem to (1) possess limited cognitive resources, leading to reliance on "rules of thumb" or "heuristics" to process information related to judgment; (2) have difficulty comprehending, manipulating, and applying numerical information, such as percentages and probabilities; and (3) be influenced greatly by contextual information that may or may not be relevant to the decision.

Use of heuristics

Blumenthal-Barby and Krieger (2015) review 19 different heuristics and biases that have been explored in medical decision-making ($N = 213$ studies). One of the most commonly studied is the availability heuristic, or reliance on accessible or otherwise memorable events in order to estimate the probability of a given event (Tversky & Kahneman, 1974). Use of this heuristic may cause people to overestimate the frequencies of relatively rare events that are particularly accessible. Given that most pathogenic genetic variants are rare, this heuristic can unwittingly exaggerate perceptions of prevalence; for example, a woman who becomes acquainted with others who have tested positive for *BRCA1/2* variants may overestimate the population prevalence of these variants, which in turn might dissuade her from being tested despite the low likelihood of testing positive even with a significant family history. In general, people overweight individual cases and underweight overall base rates—termed the "numerator bias." For example, they may consider a variant that affects 10 out of 100 individuals to be more common than a variant affecting 1 out of 10 individuals (Denes-Raj & Epstein, 1994).

A second primary heuristic used in judgment is the representativeness heuristic, a rule of thumb that largely governs judgments of similarity (Tversky & Kahneman, 1974). An example of using this heuristic might be coming to believe that a target person is likely to have a genetic variant because he or she shares attributes with someone else who has been confirmed to have the variant, without taking account of the base rate of this variant (which could be small). In addition to this tendency to underweight base rates, misuse of the representativeness heuristic undergirds several other biases. For example, people confuse conditional probabilities—believing, for example, that the probability of having Lynch syndrome given the presence of a family history of colon cancer is equivalent to the reverse probability (having a family history given Lynch syndrome) even though the latter is more probable.

Trouble with numbers

The misuse of heuristics often exemplifies people's general misunderstanding and misuse of numerical information. Several studies show that most people—even those at higher educational levels—possess only a rudimentary understanding of probabilities and other types of numerical information (Nelson, Reyna, Fagerlin, Lipkus, & Peters, 2008). For example, Lipkus, Samsa, and Rimer (2001) found in a relatively educated sample that only 51% could estimate the probability of a six-sided die coming up even, and 20% could not indicate whether 1/10, 1/100, or 1/1000 represented the highest risk. People often tend to overestimate small risks and generally have difficulty working with and comprehending very small numbers (Lichtenstein, Slovic, Fischhoff, Layman, & Combs, 1978). As a result, people may fail to understand that large relative risks can nevertheless represent small differences among small risks (e.g., a genetic variant that increases one's disease risk from 1% to 2% represents a relative risk of 100%). Thus, one cannot be confident that a recipient of genome sequencing results can fully comprehend the meaning of statistical information accompanying those results—a nontrivial concern given that many pathogenic genetic variants are rare and confer very low risk. Using verbal labels in lieu of numerical information is also problematic because people assign much higher probabilities to labels than intended by risk communicators. Berry, Knapp, and Raynor (2002) found, for example, that when the European Union elected to use the term "rare" to signify that side effects occurred in 0.01%−0.1% of users, laypeople interpreted the term as meaning the side effects occurred in about 8% of users. Cameron, Sherman, Marteau, and Brown (2009) suggest that most people think about risk as being "low" (<20%), "moderate" (20%−80%), or "high" (>80%) and fail to make finer gradations—consistent with the notion that they simply care about the "gist" of risk information (Reyna, 2004). These findings suggest the need to supply both numerical and verbal representations of risk when conveying genome sequencing results or rates of false positives and negatives.

Sensitivity to context

Contextual factors—even those unrelated to a decision—can greatly influence how information is construed, an important consideration in the communication of genetic risk information. For example, Morrison, Henderson, Taylor, A'Ch Dafydd, and Unwin (2010) presented a sample of 84 women with both positive and negative information about the value of predictive genetic testing for breast cancer. However, they manipulated the order of the information, and observed a primacy effect—the information that came first (whether negative or positive) had the greatest effect on risk perceptions, attitudes toward genetic testing, and perceived disadvantages of testing. Other work shows that even when people are informed about such order effects, they do not believe they were personally influenced by order (Nisbett & Wilson, 1977), suggesting that this effect can be particularly subtle and that genetic counselors must be cognizant of how the order of presenting genome sequencing results can affect the decision-making of their patients. People also have preferences about ordering; Loewenstein and Prelec (1993) observed, for example, that although people generally prefer to receive positive information such as test results immediately, they would rather wait to receive it after any negative information. This is notable given that genomic information can be sought out over time rather than all at once (Yu, Jamal, Tabor, & Bamshad, 2013). Another example of the importance of context is that people are also less likely to comprehend risk information when it is complex (Zikmund-Fisher, Fagerlin, & Ubel, 2010).

Recommended approaches

The literature on how cognitive limitations influence the use and comprehension of numerical information has led to several published guidelines for communicating risk information that are easily adapted to the genome sequencing setting (Fagerlin, Zikmund-Fisher, & Ubel, 2011; Fischhoff, 2012; Klein, & Ellis, in, press; Lipkus, 2007; Waters, Fagerlin, & Zikmund-Fisher, 2016; Waters, McQueen, & Cameron, 2014). These include presenting both absolute and comparative risk, avoiding the use of relative risk and "1 in X" risk formats, presenting natural frequencies in lieu of or in addition to percentages/probabilities, using both verbal and numerical risk information, and making use of pictorial representations of risk such as icon arrays and risk ladders. Some emerging evidence suggests that pictographs are even more effective when using anthropomorphic icons (people rather than dots, ovals, or other impersonal icons) and tend to be particularly helpful when conveying risk to people low in numeracy (Garcia-Retamero, Okan, & Cokely, 2012; Kreuzmair, Siegrist, & Keller, 2017). Of note, as genomic risk communication is scaled up to the population level, these recommendations are easily adapted for web-based communication protocols regarding genomic risk (Fenton et al., 2018).

Emotional influences

Accompanying the context within which information is presented are intrapersonal factors unique to the information recipient, such as emotions and motivations. In some cases, these factors can have a substantial influence. Consider a recent study by Hellwig et al. (2018) in which participants in a genome sequencing trial were presented with three hypothetical results regarding variants linked to cardiac disease that varied in severity. As expected, participants were able to distinguish the three different variants according to their level of severity, presumably because they could be easily compared. However, when Hellwig and colleagues analyzed responses to just the first variant presented (i.e., before participants had viewed the other two variants), participants' judgments depended more on their emotional reaction to the information about the variant. The authors reasoned that the lack of context imposed by only having seen one test result (without the other two to serve as context) led people to depend more on their emotional reactions.

Emotional experiences can influence judgment and decision-making in many ways. People often think about risk in more emotional terms than cognitive terms, as represented in the "affect heuristic" (Slovic et al., 2007), risk-as-feeling (Loewenstein, Weber, Hsee, & Welch, 2001), and TRIRISK (Ferrer, Klein, Persoskie, Avishai-Yitshak, & Sheeran, 2016) frameworks for risk perception. Indeed, affective risk perception—often defined as worry about experiencing an event in the future—is often more predictive of behavior than more conventional subjective likelihood measures (Dillard, Ferrer, Ubel, & Fagerlin, 2012; Janssen, Waters, van Osch, Lechner, & de Vries, 2014). Perhaps even more interestingly, affective and more logic-based risk perceptions can interact, and sometimes paradoxically. Several studies show that high worry accompanied by high perceptions of likelihood are associated with *lower* intentions to reduce one's risk, perhaps because this combination is paralyzing (Ferrer, Portnoy, & Klein, 2013; Persoskie, Ferrer, & Klein, 2014). These findings are important because they suggest that communicating high risk to someone who is already high in worry can have unintended effects.

Worry is, of course, an emotion that is directly tied to the outcome in question; for example, people may worry about the outcome of genome sequencing, which in turn might dissuade them from seeking the results (Ferrer et al., 2015). However, emotions are often more incidental to the decision context and yet still exert influence over the decision. For example, in a study by Persky, Ferrer, and Klein (2016) a virtual physician gave overweight African-American women information about the genetic underpinnings of obesity. Prior to getting the information, they viewed an unrelated film clip that made them angry. Relative to a control group, women in this anger condition revealed nonverbal behaviors consistent with poorer patient–provider communication, despite the anger being completely incidental to the interaction. Another study showed that people in an angry or fearful incidental state were less likely to engage in informed consent in the context of a clinical trial (Ferrer, Stanley, et al., 2016), a finding that could easily translate to a context in which people are engaging in genome sequencing. Emotions may also cause individuals to avoid making decisions altogether (Anderson, 2003).

People do not only experience emotion but also *anticipate* emotion. For example, Weinstein et al. (2007) found that the best predictor of whether people got a flu shot was not whether they perceived high risk for getting the flu but rather the regret they anticipated feeling in the future if they neglected to get vaccinated and then contracted the flu. Accordingly, much research suggests that people go out of their way to avoid future regret, predicting in advance that they will experience it (Gilbert & Ebert, 2002; Larrick & Boles, 1995; Simonson, 1992). Anticipated regret is only one type of anticipated emotion that can drive decisions. Ferrer et al. (2015) measured how worried participants in a genome sequencing study were about getting results suggesting they had variants that increased their risk, and also measured their predictions of how devastated they would be in the future if they learned they had such variants (anticipated affect). Consistent with Weinstein et al.'s (2007) findings, the best predictor of electing to receive genome sequencing results was anticipated affect, not current level of worry. Interestingly, people's predictions about their future affective responses tend to be exaggerated, as they fail to acknowledge the "psychological immune system" that most people draw on to cope with threatening information and experiences (Wilson & Gilbert, 2003). Indeed, a significant body of research demonstrates that people's affective reactions to genetic information are fairly muted, even in cases where the genetic information is consequential (Broadstock, Michie, & Marteau, 2000; Lerman, Croyle, Tercyak, & Hamann, 2002). One study showed that when people learned there had been a change in the interpretation of a genetic variant they were told to possess, they were largely unfazed (Taber et al., 2018). In general, people adapt to negative health news by finding meaning, changing priorities, and engaging more deeply in relationships (Biesecker & Erby, 2008; Taylor, 1983).

In summary, when evaluating how people engage with and respond to genomic information, it is important to understand that predictions about one's emotional responses to the information may be overestimated, and yet still drive decisions to receive information—even more so than current worry about the results. In general, affect appears more predictive than conventional risk perceptions of intentions and behavior in many health domains. Finally, it is important not to dismiss the potential effect of incidental emotions on people's engagement with genomic information; in counseling situations, capturing information about emotional state would appear to be prudent.

Motivational factors

Making decisions about seeking genomic risk information—and responding in a value-consistent manner to that information—can clearly be influenced by many of the cognitive biases discussed above, in concert with emotional experiences. A comprehensive analysis, however, necessitates consideration of the many enduring motives that are often present when people make personally significant decisions such as whether to receive genome sequencing results and how to act on them. These motives can often supersede reasoned, logical, and rational approaches to such decision-making. Consequently, designing effective risk communication and facilitating genomic decision-making must not only be a function of minimizing the effects of cognitive biases and leveraging emotional influences but also heeding the powerful role played by people's motives. In particular, people aspire to (1) conserve their cognitive resources, (2) maintain positive self-views and act consistently with their values, (3) perceive control over personally relevant outcomes, (4) regard themselves (and be regarded by others) as rational actors, (5) engage in socially normative behavior, and (6) avoid loss, uncertainty, and ambiguity. We briefly consider each of these in turn.

Conserving cognitive resources

Not only do people have limited cognitive resources given the wealth of stimuli to which they are exposed—leading to many of the heuristics and biases highlighted earlier—but people also seem implicitly aware of the limitations on their cognitive resources and thus appear motivated to protect their cognitive resources. This motive can manifest in several ways. One is a confirmation bias, such that people satisfice by paying attention only to information that confirms their beliefs than to information that disconfirms it—with the latter information often being more difficult and therefore more costly to obtain (Klayman & Ha, 1987). In general, people protect their beliefs even when faced with disconfirming information. For example, an individual might believe he is genetically predisposed to addiction based on prior experience, and thus disbelieve genomic testing results suggesting the reverse (see Dillard, McCaul, Kelso, & Klein [2006] for a similar example involving trust in Gail model cancer risk estimates). Another example of protecting cognitive resources is making attempts to reduce the number and complexity of choices given the uncomfortable state of "choice overload" (Iyengar & Lepper, 2000; Schwartz, 2004). People might be faced with a choice of multiple genome sequencing results at once, several results spaced out over time, or results "binned" into intuitively comprehensible disease categories and may elect for the latter to reduce complexity. They may also accede willingly to defaults, making the assumption that defaults represent what experts (or peers) consider to be the "best" or most normative decision. Indeed, defaults provide powerful cues to action (Johnson & Goldstein, 2003). Although defaults are appropriate in some health contexts—such as designing a building to make stair use rather than elevators a default method of mobility in the building—the use of defaults may be more ethically questionable in the context of genome sequencing where choices must be personally driven and value-sensitive.

Maintaining positive self-views and affirming one's values

People are generally motivated to view their abilities, relationships, values, and personality attributes in a positive manner, and in fact often hold self-beliefs that are illusory relative to objective standards (Dunning, Heath, & Suls, 2018). For example, people tend to be unrealistically optimistic about their risk of experiencing future health outcomes (Weinstein, 1980). That is more the case for controllable outcomes (Harris, 1996), an important caveat given that genomic test results are sometimes not actionable. When people's positively biased beliefs are threatened, they often engage in many strategies designed to sustain those beliefs (e.g., McQueen, Vernon, & Swank, 2013). People also actively avoid information that may be helpful but potentially threatening to the self (Howell & Shepperd, 2012). One study found that people who tend to avoid threatening information in general are also less likely to seek out the results of their genome sequencing (Taber, Klein, Ferrer, Lewis, Harris, et al., 2015).

If people underestimate their risk of health outcomes, they may be less likely to seek potentially diagnostic genetic information from genome sequencing, suggesting the need to minimize the effects of unrealistically optimistic risk perceptions. Reducing the bias itself may not be the most prudent approach, as positive illusions have been linked in some studies to beneficial mental and physical health outcomes (Persoskie, Ferrer, Nelson, & Klein, 2014; Taylor & Brown, 1994). A more effective approach may be to take advantage of another self-related motive, which is to see oneself as acting consistently with one's values. One study showed that when people who were unrealistically optimistic about their cancer risk had a chance to talk about positive aspects of their health prior to receiving a persuasive message about cancer screening, they were more likely to intend to screen (Klein et al., 2010). Taber, Klein, Ferrer, Lewis, Harris, et al. (2015) observed that people who tend to focus on values in the context of daily threats were less likely to exhibit the relationship noted above between information avoidance tendencies and disinterest in receiving genome sequencing results. Several studies show that giving people a chance to reflect on their values prior to the delivery of potentially threatening information reduces avoidance of and defensive reactions toward that information (for reviews, see Epton, Harris, Kane, van Koningsbruggen, & Sheeran, 2015; Sweeney & Moyer, 2015). The promise of such "self-affirmation" interventions has not been systematically explored in genetic testing contexts.

Perceiving control over personally relevant outcomes

People tend to desire control over personally relevant outcomes (Langer, 1975). Perceptions that one has little control can promote helplessness, anxiety, and depression so much so that having an "illusion of control" has been argued to be a pillar of good mental health (Taylor & Brown, 1988). Consistent with the self-judgment biases noted earlier, people also believe they are better than others at exerting control over important outcomes (Klein & Kunda, 1994). Moreover, the endorsement of the belief that important outcomes are controllable can lead to a belief in a just world and victim blaming (Haynes & Olson, 2006; Lerner, 1980). In the genetic testing and genome sequencing context, the motive to perceive and maintain control may exercise influence in several ways. For example, people are likely to want control over the accessibility and use of their genomic information, to desire genome sequencing results that are actionable more than those that are nonactionable, and to hope that one's actions can reduce any debilitating effects of genetic disorders. Perhaps most importantly, providing opportunities for people to exert control can promote interest in taking self-protective actions and lead to desirable outcomes. In one classic study, Langer and Rodin (1976) observed beneficial health effects among nursing home residents who were simply given the responsibility of taking care of a plant or greater choice in their entertainment options.

An early concern raised about the prospect of returning genetic test results was that recipients would come to attribute disease entirely to uncontrollable genetic causes and therefore express less interest in adopting lifestyle behaviors to reduce risk (Hunter, Khoury, & Drazen, 2008). Subsequent findings seem to ameliorate this concern, however. People generally appear to understand that in most cases genetics interact with behavior (and other causes) to produce disease, perhaps acknowledging even more the importance of healthy behaviors that might offset one's genetic risk (Kaphingst et al., 2012).

Regarding oneself (and being regarded by others) as a rational actor

In general, people aspire to see themselves (and to be seen by others) as individuals who make rational, defensible decisions based on reason (Shafir, Simonson, & Tversky, 1993). Paradoxically, this motive can lead to judgments and decisions that in reality violate the assumptions of prescriptive decision theories and potentially lead to poorer decision-making. One example is the compromise effect, or the related attraction effect (Simonson, 1989). Suppose a patient is presented with two drugs, A and B, such that A is safer and B is more effective. Suppose further that the patient values safety and effectiveness equally, making the choice between A and B comparable to a state of equipoise. If the patient was then presented with a third option, C, which was slightly less safe and slightly less effective than B, then B might become more appealing as it dominates C on both of the important characteristics (safety and effectiveness). Choosing B would exemplify the attraction effect. Alternatively, if the new option C is even more effective than B but less safe, B might then be viewed as an acceptable compromise given that, relative to A and C, it possesses moderate safety and effectiveness. Choosing B would exemplify the compromise effect. In both cases, the choice appears to be driven by an attempt to make a rational, defensible decision. Compromises are usually perceived as rational outcomes in a difficult decision context.

Given that people might be faced with many different options with respect to genetic testing—e.g., single gene tests vs. multiplex testing vs. genome sequencing—one must consider how the need to appear rational can drive patients to prefer one approach over another.

Engaging in normative behavior

People tend to care deeply how they compare with others on many dimensions such as ability, morality, intelligence, and appearance, prompting Festinger (1954) to label social comparison a drive much like thirst and hunger. Social comparisons are often automatic (Gilbert, Giesler, & Morris, 1995), which is not surprising given there is evolutionary significance for a conspecific knowing his or her standing relative to others in the species. Relatedly, people are particularly sensitive to social norms and in general attempt to adhere to such norms given the social consequences of violating them (Christensen, Rothgerber, Wood, & Matz, 2004; Hechter & Opp, 2001). Not surprisingly, then, many health judgments and decisions are influenced by social comparisons and social norms (for review, see Klein & Rice, in press). Taber, Klein, Ferrer, Lewis, Biesecker, et al. (2015) observed that comparative risk perceptions (i.e., how one's risk compares with that of other people) regarding the chances of having a pathogenic genetic variant were more predictive of intentions to get genome sequencing results than other types of risk perceptions (though only among individuals who were dispositionally optimistic). Facio et al. (2012) found that perceptions of social norms regarding interest in genome sequencing results were high, and Reid et al. (2018) found that these norms were predictive of decisions to obtain such results. An implication of these findings is that genetic counselors might present social norms as a reference point when helping clients make decisions about whether to be tested.

Avoiding loss, uncertainty, and ambiguity

People find losses more painful than they find gains appealing, following prospect theory (Kahneman & Tversky, 2013). Consequently, when faced with possible losses, people are more risk-seeking; that is, they are more willing to embrace risk in the hope that doing so will minimize loss and the pain associated with loss. Accordingly, it is easier to encourage people to engage in risky behaviors such as screening by using messages that focus on the loss of not screening rather than the gains of screening (Rothman, Kelly, Hertel, & Salovey, 2003). In cases where genetic testing is beneficial, loss-based messages may be more effective at encouraging such testing (given that testing, like screening, is imbued with the risk of getting bad news). Of course, in many cases it is unclear whether obtaining genetic information is beneficial, highlighting the importance of neutrality when working with patients. Importantly, however, genetic counselors need to be aware that talking about the benefits of genetic testing with either a gain frame or a loss frame might unwittingly nudge a client in one direction or another (Barr, 2015). Many of the findings discussed in this chapter suggest that it is possible to unintentionally nudge a client in one direction or another based on how information is conveyed and framed, highlighting the importance of training in decision science for providers delivering genetic information.

In addition to loss, people also generally abhor uncertainty and ambiguity (defined as uncertainty about uncertainty, such as exposure to a range of probabilities rather than a point estimate; Han et al., 2017; Han, Klein, & Arora, 2011). Genome sequencing results are often imbued with both uncertainty and ambiguity in that it is unclear whether a particular variant will lead to a disease (uncertainty), and the chances that it does may vary greatly based on other characteristics (ambiguity). This aversion can have consequences for judgment and decision-making in medical contexts. For example, Geller, Tambor, Chase, and Holtzman (1993) found that physicians who were relatively lower on an individual difference measure called "tolerance for ambiguity" were more likely to withhold genetic test results and encourage patients to terminate their pregnancies. Taber, Klein, Ferrer, Han, et al. (2015) found that people who perceived genome sequencing results to be more ambiguous also harbored less favorable beliefs about genome sequencing results as well as lower intentions to learn their results and share them with family members. Biesecker et al. (2017) developed a measure called the Perceptions of Uncertainties in Genome Sequencing scale, which systematically examines numerous types of uncertainty (clinical, evaluative, and affective) that people might experience in the genome sequencing context. Respondents are asked, for example, the extent to which they are uncertain about "whether I will be reassured or encouraged by my future test results" and "whether I will be able to trust my future test results." The authors found that people scoring high on perceived uncertainty expressed more ambivalence about receiving their genome sequencing results.

Summary

People bring a variety of core motives to the genetic testing setting, many of which can have profound effects on decisions made in that setting. When making decisions, people generally lean toward preserving their cognitive resources, acting normatively, maintaining positive views of themselves—including the perception that they make rational, defensive decisions— and sustaining perceptions of control over personal outcomes while avoiding loss, uncertainty, and ambiguity. It is essential, then, to heed the hovering effects of these motives in communications about the value of genome sequencing and also when delivering results.

CONCLUSION

To truly understand how people engage with and respond to genomic information, it is crucial to appreciate the psychology of judgment and decision-making.

Clinicians might often be perplexed by the many factors that contribute to client perceptions and decisions, and knowledge of the relevant literature may be particularly helpful. In general, clinicians best not presume that providing more information leads to more accurate risk perceptions or more informed health decisions. Rather, the expectation should be that the information is not only conceptually complex and often unfamiliar, but also demands numeracy skills to understand probability and relatively high health literacy. Further, the threatening nature of genetic information leads patients to reflexively engage responses to buffer that threat. Such attempts to deflect can thwart opportunities to seek preventive or screening options and suggest the need for deliberate efforts to circumvent these barriers. Practically, it may be difficult to heed the many factors reviewed in this chapter in every client interaction, but attention to even one or two (e.g., simplifying the presentation of numerical information, providing social norms, presenting choices in such a way as to avoid order effects) could make an important difference in a subsequent decision.

Although not all of the factors discussed in this chapter can be routinely considered in genetic counseling and decision-making, such as the nondelibrate use of intuitive heuristics, many can be addressed. If practitioners are familiar with the potential consequences of their approaches and actively work to circumvent the effects of these approaches, patients may be activated to make better decisions. In the future, evidence-based interactive platforms designed as decision tools may include simplified multiple formats to present essential information and facilitate informed decision-making that avoids some of the barriers discussed in this chapter. Significant research endeavors will be needed to assess the effectiveness of interactive decision tools. In the meantime, we advocate for providers to appreciate the important role of cognitive, affective, and motivating factors among their patients and work to promote accurate understanding of genetic risks and their options to manage or avert those risks.

REFERENCES

Anderson, C. J. (2003). The psychology of doing nothing: Forms of decision avoidance result from reason and emotion. *Psychological Bulletin, 129*(1), 139–167.

Barr, M. L. (2015). Testing for hereditary cancer predisposition: The impact of the number of options and a provider recommendation on decision-making outcomes (Master's thesis). Johns Hopkins Bloomberg School of Public Health.

Berry, D. C., Knapp, P., & Raynor, D. (2002). Provision of information about drug side-effects to patients. *Lancet, 359*(9309), 853–854.

Biesecker, B. B., & Erby, L. (2008). Adaptation to living with a genetic condition or risk: A mini-review. *Clinical Genetics, 74*(5), 401–407.

Biesecker, B. B., Woolford, S. W., Klein, W. M. P., Brothers, K. B., Umstead, K. L., Lewis, K. L., . . . Han, P. K. (2017). PUGS: A novel scale to assess perceptions of uncertainties in genome sequencing. *Clinical Genetics, 92*(2), 172–179.

Biesecker, L. G., & Green, R. C. (2014). Diagnostic clinical genome and exome sequencing. *New England Journal of Medicine, 370*(25), 2418–2425.

Blumenthal-Barby, J. S., & Krieger, H. (2015). Cognitive biases and heuristics in medical decision making: A critical review using a systematic search strategy. *Medical Decision Making, 35*(4), 539−557.

Broadstock, M., Michie, S., & Marteau, T. (2000). Psychological consequences of predictive genetic testing: A systematic review. *European Journal of Human Genetics, 8*(10), 731−738.

Cameron, L. D., Sherman, K. A., Marteau, T. M., & Brown, P. M. (2009). Impact of genetic risk information and type of disease on perceived risk, anticipated affect, and expected consequences of genetic tests. *Health Psychology, 28*(3), 307−316.

Christensen, P. N., Rothgerber, H., Wood, W., & Matz, D. C. (2004). Social norms and identity relevance: A motivational approach to normative behavior. *Personality and Social Psychology Bulletin, 30* (10), 1295−1309.

Denes-Raj, V., & Epstein, S. (1994). Conflict between intuitive and rational processing: When people behave against their better judgment. *Journal of Personality and Social Psychology, 66*(5), 819−829.

Dillard, A. J., Ferrer, R. A., Ubel, P. A., & Fagerlin, A. (2012). Risk perception measures' associations with behavior intentions, affect, and cognition following colon cancer screening messages. *Health Psychology, 31*(1), 106−113.

Dillard, A. J., McCaul, K. D., Kelso, P. D., & Klein, W. M. P. (2006). Resisting good news: Reactions to breast cancer risk communication. *Health Communication, 19*(2), 115−123.

Dunning, D., Heath, C., & Suls, J. M. (2018). Reflections on self-reflection: Contemplating flawed self-judgments in the clinic, classroom, and office cubicle. *Perspectives on Psychological Science, 13*(2), 185−189.

Epton, T., Harris, P. R., Kane, R., van Koningsbruggen, G. M., & Sheeran, P. (2015). The impact of self-affirmation on health-behavior change: A meta-analysis. *Health Psychology, 34*(3), 187−196.

Facio, F. M., Eidem, H., Fisher, T., Brooks, S., Linn, A., Kaphingst, K. A., ... Biesecker, B. B. (2012). Intentions to receive individual results from whole-genome sequencing among participants in the ClinSeq study. *European Journal of Human Genetics, 21*(3), 261−265.

Fagerlin, A., Zikmund-Fisher, B. J., & Ubel, P. A. (2011). Helping patients decide: Ten steps to better risk communication. *Journal of the National Cancer Institute, 103*(19), 1436−1443.

Fenton, G. L., Smit, A. K., Freeman, L., Badcock, C., Dunlop, K., Butow, P. N., ... Cust, A. E. (2018). Development and evaluation of a telephone communication protocol for the delivery of personalized melanoma genomic risk to the general population. *Journal of Genetic Counseling, 27*(2), 370−380.

Ferrer, R. A., Klein, W. M. P., Persoskie, A., Avishai-Yitshak, A., & Sheeran, P. (2016). The tripartite model of risk perception (TRIRISK): Distinguishing deliberative, affective, and experiential components of perceived risk. *Annals of Behavioral Medicine, 50*(5), 653−663.

Ferrer, R. A., Portnoy, D. B., & Klein, W. M. P. (2013). Worry and risk perceptions as independent and interacting predictors of health protective behaviors. *Journal of Health Communication, 18*(4), 397−409.

Ferrer, R. A., Stanley, J. T., Graff, K., Klein, W. M. P., Goodman, N., Nelson, W. L., & Salazar, S. (2016). The influence of emotion on the informed consent process in cancer clinical trials. *Journal of Behavioral Decision-Making, 29*, 245−253.

Ferrer, R. A., Taber, J. M., Klein, W. M. P., Harris, P. R., Lewis, K. L., & Biesecker, L. G. (2015). The role of current affect, anticipated affect and spontaneous self-affirmation in decisions to receive self-threatening genetic risk information. *Cognition and Emotion, 29*(8), 1456−1465.

Festinger, L. (1954). A theory of social comparison processes. *Human Relations, 7*(2), 117−140.

Fischhoff, B. (2012). *Communicating risks and benefits: An evidence-based user's guide.* Silver Spring, Maryland: Government Printing Office.

Garcia-Retamero, R., Okan, Y., & Cokely, E. T. (2012). Using visual aids to improve communication of risks about health: A review. *The Scientific World Journal, 2012*, 1−10.

Geller, G., Tambor, E. S., Chase, G. A., & Holtzman, N. A. (1993). Measuring physicians' tolerance for ambiguity and its relationship to their reported practices regarding genetic testing. *Medical Care,* 989−1001.

Gilbert, D. T., & Ebert, J. E. (2002). Decisions and revisions: The affective forecasting of changeable outcomes. *Journal of Personality and Social Psychology, 82*(4), 503−514.

Gilbert, D. T., Giesler, R. B., & Morris, K. A. (1995). When comparisons arise. *Journal of Personality and Social Psychology, 69*(2), 227−236.

Gooding, H. C., Organista, K., Burack, J., & Biesecker, B. B. (2006). Genetic susceptibility testing from a stress and coping perspective. *Social Science and Medicine, 62*(8), 1880–1890.

Han, P. K., Klein, W. M. P., & Arora, N. K. (2011). Varieties of uncertainty in health care: A conceptual taxonomy. *Medical Decision Making, 31*(6), 828–838.

Han, P. K., Umstead, K. L., Bernhardt, B. A., Green, R. C., Joffe, S., Koenig, B., ... Biesecker, B. B. (2017). A taxonomy of medical uncertainties in clinical genome sequencing. *Genetics in Medicine, 19* (8), 918–925.

Harris, P. R. (1996). Sufficient grounds for optimism?: The relationship between perceived controllability and optimistic bias. *Journal of Social and Clinical Psychology, 15*(1), 9–52.

Haynes, G. A., & Olson, J. M. (2006). Coping with threats to just-world beliefs: Derogate, blame, or help? *Journal of Applied Social Psychology, 36*(3), 664–682.

Hechter, M., & Opp, K.-D. (2001). *Social norms*. New York: Russell Sage Foundation.

Hellwig, L. D., Biesecker, B. B., Lewis, K. L., Biesecker, L. G., James, C. A., & Klein, W. M. P. (2018). Ability of patients to distinguish among cardiac genomic variant subclassifications. *Circulation: Genomic and Precision Medicine, 11*(6), e001975.

Howell, J. L., & Shepperd, J. A. (2012). Behavioral obligation and information avoidance. *Annals of Behavioral Medicine, 45*(2), 258–263.

Hunter, D. J., Khoury, M. J., & Drazen, J. M. (2008). Letting the genome out of the bottle—Will we get our wish? *New England Journal of Medicine, 358*(2), 105–107.

Iyengar, S. S., & Lepper, M. R. (2000). When choice is demotivating: Can one desire too much of a good thing? *Journal of Personality and Social Psychology, 79*(6), 995–1006.

Janssen, E., Waters, E. A., van Osch, L., Lechner, L., & de Vries, H. (2014). The importance of affectively-laden beliefs about health risks: The case of tobacco use and sun protection. *Journal of Behavioral Medicine, 37*(1), 11–21.

Johnson, E. J., & Goldstein, D. (2003). Do defaults save lives? *Science, 302*(5649), 1338–1339.

Kahneman, D., & Tversky, A. (2013). Prospect theory: An analysis of decision under risk. In L. MacLean, & W. Ziemba (Eds.), *Handbook of the fundamentals of financial decision making: Part I* (pp. 99–127). World Scientific.

Kaphingst, K. A., McBride, C. M., Wade, C., Alford, S. H., Reid, R., Larson, E., ... Brody, L. C. (2012). Patients' understanding of and responses to multiplex genetic susceptibility test results. *Genetics in Medicine, 14*(7), 681–687.

Khan, C. M., Rini, C., Bernhardt, B. A., Roberts, J. S., Christensen, K. D., Evans, J. P., ... Henderson, G. E. (2015). How can psychological science inform research about genetic counseling for clinical genomic sequencing? *Journal of Genetic Counseling, 24*(2), 193–204.

Klayman, J., & Ha, Y.-W. (1987). Confirmation, disconfirmation, and information in hypothesis testing. *Psychological Review, 94*(2), 211–228.

Klein, W. M., & Kunda, Z. (1994). Exaggerated self-assessments and the preference for controllable risks. *Organizational Behavior and Human Decision Processes, 59*(3), 410–427.

Klein, W. M. P., & Ellis, E. M. (in press). Effective and impactful risk communication. To appear In M. Boulton & R. Wallace (Eds.), *Maxcy-Rosenau-Last public health and preventive medicine* (16th ed.). New York: McGraw-Hill.

Klein, W. M. P., Lipkus, I. M., Scholl, S. M., McQueen, A., Cerully, J. L., & Harris, P. R. (2010). Self-affirmation moderates effects of unrealistic optimism and pessimism on reactions to tailored risk feedback. *Psychology and Health, 25*(10), 1195–1208.

Klein, W.M.P., & Rice, E. (in press). Health cognitions, decision-making and behavior: The ubiquity of social comparison. To appear In J. Suls, R. L. Collins, & L. Wheeler (Eds.). *Social comparison in judgment and behavior*. New York: Oxford University Press.

Kreuzmair, C., Siegrist, M., & Keller, C. (2017). Does iconicity in pictographs matter? The influence of iconicity and numeracy on information processing, decision making, and liking in an eye-eracking study. *Risk Analysis, 37*(3), 546–556.

Langer, E. J. (1975). The illusion of control. *Journal of Personality and Social Psychology, 32*(2), 311–328.

Langer, E. J., & Rodin, J. (1976). The effects of choice and enhanced personal responsibility for the aged: A field experiment in an institutional setting. *Journal of Personality and Social Psychology, 34*(2), 191–198.

Larrick, R. P., & Boles, T. L. (1995). Avoiding regret in decisions with feedback: A negotiation example. *Organizational Behavior and Human Decision Processes*, 87−97.

Lerman, C., Croyle, R. T., Tercyak, K. P., & Hamann, H. (2002). Genetic testing: Psychological aspects and implications. *Journal of Consulting and Clinical psychology*, 70(3), 784−797.

Lerner, M. J. (1980). The belief in a just world. In M. J. Lerner (Ed.), *The Belief in a Just World. Perspectives in Social Psychology* (pp. 9−30). Boston, MA: Springer.

Lichtenstein, S., Slovic, P., Fischhoff, B., Layman, M., & Combs, B. (1978). Judged frequency of lethal events. *Journal of Experimental Psychology: Human Learning Memory*, 4(6), 551−578.

Lipkus, I. M. (2007). Numeric, verbal, and visual formats of conveying health risks: Suggested best practices and future recommendations. *Medical Decision Making*, 27(5), 696−713.

Lipkus, I. M., Samsa, G., & Rimer, B. K. (2001). General performance on a numeracy scale among highly educated samples. *Medical Decision Making*, 21(1), 37−44.

Loewenstein, G. F., & Prelec, D. (1993). Preferences for sequences of outcomes. *Psychological Review*, 100 (1), 91−108.

Loewenstein, G. F., Weber, E. U., Hsee, C. K., & Welch, N. (2001). Risk as feelings. *Psychological Bulletin*, 127(2), 267−286.

Marteau, T. M., Dormandy, E., & Michie, S. (2001). A measure of informed choice. *Health Expectations*, 4(2), 99−108.

McQueen, A., Vernon, S. W., & Swank, P. R. (2013). Construct definition and scale development for defensive information processing: An application to colorectal cancer screening. *Health Psychology*, 32 (2), 190−202.

Morrison, V., Henderson, B. J., Taylor, C., A'Ch Dafydd, N., & Unwin, A. (2010). The impact of information order on intentions to undergo predictive genetic testing: An experimental study. *Journal of Health Psychology*, 15(7), 1082−1092.

Nelson, W., Reyna, V. F., Fagerlin, A., Lipkus, I., & Peters, E. (2008). Clinical implications of numeracy: Theory and practice. *Annals of Behavioral Medicine*, 35(3), 261−274.

Nisbett, R. E., & Wilson, T. D. (1977). Telling more than we can know: Verbal reports on mental processes. *Psychological Review*, 84(3), 231−259.

Persky, S., Ferrer, R. A., & Klein, W. M. P. (2016). Nonverbal and paraverbal behavior in (simulated) medical visits related to genomics and weight: A role for emotion and race. *Journal of Behavioral Medicine*, 39(5), 804−814.

Persoskie, A., Ferrer, R. A., & Klein, W. M. P. (2014). Association of cancer worry and perceived risk with doctor avoidance: An analysis of information avoidance in a nationally representative US sample. *Journal of Behavioral Medicine*, 37(5), 977−987.

Persoskie, A., Ferrer, R. A., Nelson, W. L., & Klein, W. M. P. (2014). Precancer risk perceptions predict postcancer subjective well-being. *Health Psychology*, 33(9), 1023−1032.

Reid, A. E., Taber, J. M., Ferrer, R. A., Biesecker, B. B., Lewis, K. L., Biesecker, L. G., & Klein, W. M. P. (2018). Associations of perceived norms with intentions to learn genomic sequencing results: Roles for attitudes and ambivalence. *Health Psychology*, 37(6), 553−561.

Reyna, V. F. (2004). How people make decisions that involve risk: A dual-processes approach. *Current Directions in Psychological Science*, 13(2), 60−66.

Rothman, A. J., Kelly, K. M., Hertel, A. W., & Salovey, P. (2003). Message frames and illness representations: Implications for interventions to promote and sustain healthy behavior. In L. Cameron, & H. Leventhal (Eds.), *The self-regulation of health and illness behaviour* (pp. 278−296). New York: Routledge.

Schwartz, B. (2004). *The paradox of choice: Why more is less*. New York: Ecco.

Shafir, E., & LeBoeuf, R. A. (2002). Rationality. *Annual Review of Psychology*, 53(1), 491−517.

Shafir, E., Simonson, I., & Tversky, A. (1993). Reason-based choice. *Cognition*, 49(1), 11−36.

Simonson, I. (1989). Choice based on reasons: The case of attraction and compromise effects. *Journal of Consumer Research*, 16(2), 158−174.

Simonson, I. (1992). The influence of anticipating regret and responsibility on purchase decisions. *Journal of Consumer Research*, 19(1), 105−118.

Skinner, D., Roche, M. I., Weck, K. E., Raspberry, K. A., Foreman, A. K. M., Strande, N. T., ... Henderson, G. E. (2018). "Possibly positive or certainly uncertain?": Participants' responses to uncertain diagnostic results from exome sequencing. *Genetics in Medicine*, 20(3), 313−319.

Slovic, P., Finucane, M. L., Peters, E., & MacGregor, D. G. (2007). The affect heuristic. *European Journal of Operational Research, 177*(3), 1333—1352.

Steele, C. M. (1988). The psychology of self-affirmation: Sustaining the integrity of the self. *Advances in Experimental Social Psychology, 21,* 261—302.

Sweeney, A. M., & Moyer, A. (2015). Self-affirmation and responses to health messages: A meta-analysis on intentions and behavior. *Health Psychology, 34*(2), 149—159.

Taber, J. M., Klein, W. M. P., Ferrer, R. A., Han, P. K., Lewis, K. L., Biesecker, L. G., & Biesecker, B. B. (2015). Perceived ambiguity as a barrier to intentions to learn genome sequencing results. *Journal of Behavioural Medicine, 38*(5), 715—726.

Taber, J. M., Klein, W. M. P., Ferrer, R. A., Lewis, K. L., Biesecker, L. G., & Biesecker, B. B. (2015). Dispositional optimism and perceived risk interact to predict intentions to learn genome sequencing results. *Health Psychology, 34*(7), 718—728.

Taber, J. M., Klein, W. M. P., Ferrer, R. A., Lewis, K. L., Harris, P. R., Shepperd, J. A., & Biesecker, L. G. (2015). Information avoidance tendencies, threat management resources, and interest in genetic sequencing feedback. *Annals of Behavioral Medicine, 49*(4), 616—621.

Taber, J. M., Klein, W. M. P., Lewis, K. L., Johnston, J. J., Biesecker, L. G., & Biesecker, B. B. (2018). Reactions to clinical reinterpretation of a gene variant by participants in a sequencing study. *Genetics in Medicine, 20*(3), 337—345.

Taylor, S. E. (1983). Adjustment to threatening events: A theory of cognitive adaptation. *American Psychologist, 38*(11), 1161—1173.

Taylor, S. E., & Brown, J. D. (1988). Illusion and well-being: A social psychological perspective on mental health. *Psychological Bulletin, 103*(2), 193—210.

Taylor, S. E., & Brown, J. D. (1994). Positive illusions and well-being revisited: Separating fact from fiction. *Psychological Bulletin, 116*(1), 21—27.

Tversky, A., & Kahneman, D. (1974). Judgment under uncertainty: Heuristics and biases. *Science, 185* (4157), 1124—1131.

Waters, E. A., Fagerlin, A., & Zikmund-Fisher, B. J. (2016). Overcoming the many pitfalls of communicating risk. In M. Diefenbach, S. Miller-Halegoua, & D. Bowen (Eds.), *Handbook of health decision science* (pp. 265—277). Verlag, NY: Springer.

Waters, E. A., McQueen, A., & Cameron, L. D. (2014). Perceived risk and health risk communication. In L. Martin, & R. DiMatteo (Eds.), *The Routledge handbook of language and health communication* (pp. 47—60). New York: Routledge.

Weinstein, N. D. (1980). Unrealistic optimism about future life events. *Journal of Personality and Social Psychology, 39*(5), 806—820.

Weinstein, N. D., Kwitel, A., McCaul, K. D., Magnan, R. E., Gerrard, M., & Gibbons, F. X. (2007). Risk perceptions: Assessment and relationship to influenza vaccination. *Health Psychology, 26*(2), 146—151.

Werner-Lin, A., Zaspel, L., Carlson, M., Mueller, R., Walser, S. A., Desai, R., & Bernhardt, B. A. (2018). Gratitude, protective buffering, and cognitive dissonance: How families respond to pediatric whole exome sequencing in the absence of actionable results. *American Journal of Medical Genetics Part A, 176*(3), 578—588.

Wilson, T. D., & Gilbert, D. T. (2003). Affective forecasting. *Advances in Experimental Social Psychology, 35*(35), 345—411.

Yu, J.-H., Jamal, S. M., Tabor, H. K., & Bamshad, M. J. (2013). Self-guided management of exome and whole-genome sequencing results: Changing the results return model. *Genetics in Medicine, 15*(9), 684—690.

Zikmund-Fisher, B. J., Fagerlin, A., & Ubel, P. A. (2010). A demonstration of "less can be more" in risk graphics. *Medical Decision Making, 30*(6), 661—671.

CHAPTER 5

Uncertainties in Genome Sequencing

Barbara B. Biesecker[1], Aad Tibben[2] and Joel Vos[3]
[1]Distinguished Fellow, RTI International, Bethesda, MD, United States
[2]Department of Clinical Genetics, Leiden University Medical Centre, Leiden, The Netherlands
[3]Department of Psychology, University of Roehampton, London, United Kingdom

Uncertainties pervade life and medicine; genomics is no exception. Yet, return of results from genomic sequencing greatly exceeds the frequency of uncertain results of single-gene testing and of uncertain results in medicine generally. Undergoing genomic sequencing comes with a high probability of the return of variants of unknown clinical significance (VUS). A VUS is a genetic variant whose association with disease risk is unknown yet. Existing information about a variant lacks reliability, credibility, or adequacy to classify the sequence change either as a normal variation (benign) or a disease-causing mutation and is therefore ambiguous (Han, Klein, & Arora, 2011). As such, informing patients to establish the expectation for VUS results at the time of obtaining consent, exploring with patients how they may respond to uncertain results, and helping them navigate uncertain results upon return, are all valuable but challenging aspects of implementing genomic sequencing. Overall, there is a widespread need to help research participants, patients, and the public to learn to expect and accept uncertainties from genomic tests for the foreseeable future. Additionally, individuals may benefit from differentiation of uncertainties that can be modified (aleatory) and those that cannot (epistemic). For example, the uncertainties about whether a person with a mutation that predisposes to hereditary cancer risk will develop cancer are not modifiable—it will happen or not (epistemic). In contrast, the degree of risk may be reduced by prophylactic surgery (aleatory). In one case, patients must come to manage that which is not modifiable and in the other they may effectively pursue opportunities to reduce the uncertainty. While most patients find psychological and behavioral approaches to manage both types of uncertainties, healthcare providers can expedite the process of identifying and pursuing aleatory uncertainties to mitigate negative health outcomes. Significant variation in how patients manage uncertainties relates to differences in the extent to which patients need clarity or are able to tolerate ambiguity (Han, Reeve, & Moser, 2009; Hillen, Gutheil, Strout, Smets, & Han, 2017). Personality traits such as optimism and resilience are associated with lower perception of uncertainty and higher psychological wellbeing (Taber et al., 2015).

Clinical Genome Sequencing
DOI: https://doi.org/10.1016/B978-0-12-813335-4.00005-2

HISTORY OF UNCERTAINTY IN GENETICS

Awareness of heredity is as old as mankind. The "curse" of heredity appeared in the bible. Among the first descriptions of uncertainty in the medical field are the observations of George Huntington of a chorea appearing in two generations of a New England family (Harper, 2014). Heredity was considered a social-economic threat to society in the early decades of the 20th century, eventually leading to the race laws in Germany and healthcare policies in countries such as the United States and Sweden. The discovery of DNA by Watson and Crick in the early 1950s opened the doors to opportunities to develop insight into the dynamics of the onset and course of genetic disorders, and to the location and identification of the genetic aberrations that led to manifestation of diseases. In medicine in the 1970s, clinical genetics was established as a new discipline that focused primarily on dysmorphology in newborns and descriptions of patterns of the inheritance of traits. Pedigrees of hereditary disorders with onset later in life were drawn in the 1970s and 1980s and new research tools were developed, such as linkage analysis, that allowed the demonstration of genetic susceptibility to disease. Even if genetic mutations could not be demonstrated, risk assessment based on family histories offered people with significant uncertainties about their genetic risk some relief or at least the ability to understand their disease risks and learn to manage them over time.

It is easy to forget how recently we were naïve of the mechanisms of human inheritance and how far the science has advanced in the last 50 years. In this chapter, we discuss a plethora of uncertainties in genomics, but ones that are couched in contemporary understanding of genomic science. The genome of any individual can now be sequenced, and it can be assessed for known gene variants both related to increased risks of the individual and to secondary findings that are known to incur significant health risks and are medically actionable (see Chapter 2: Genome Sequencing and Individual Responses to Results). Those of us who care for patients and research participants have an obligation to help those who undergo genome sequencing to appreciate how much of it remains unknown and that there are often changes in genes identified that are benign or pathogenic, but for which there is insufficient evidence to determine which and thus are returned as a VUS. While interpretation will improve with time, identification of VUS results is expected to continue for decades to come as significant population-level data are needed for more comprehensive interpretation of a human genome. Herein we describe the dimensions of uncertainties in genomic sequencing; and how uncertainties are communicated, perceived, and managed.

DIMENSIONS FOR UNCERTAINTIES IN GENOME SEQUENCING

Uncertainty may be defined as the "subjective perception of ignorance" that implies a conscious awareness of one's lack of knowledge (Han et al., 2011).

Biesecker et al. (2017) developed a taxonomy of uncertainties in genomic sequencing, based on the taxonomy of medical uncertainties by Han et al. (2011). Han and colleagues proposed an integrative taxonomy that categorizes uncertainty related to its sources, issues, and locus (Bloss et al., 2015; Han et al., 2011). In this model, possible sources of uncertainty include probability, ambiguity, and complexity. Probability refers to the indeterminacy of future outcomes, ambiguity refers to the lack of reliability, credibility, or adequacy of information, and complexity refers to the features of information that make it difficult to understand, such as multiplicity of risks. The taxonomy of uncertainties in genomics maintains Han's categories of sources, issues, and locus. It originated from interviews with genomics researchers, clinicians, and counselors and extraction of key domains from a literature review of uncertainties in genomics. The taxonomy aims to categorize the multidimensions and domains of uncertainties that exist related to clinical use of genomic sequencing. Notable among the many dimensions is the uncertainty in the data used by laboratory scientists to interpret variants. Often there is insufficient evidence for the consequences of a variant to interpret it as pathogenic or benign. This is a major source of uncertain information that may need to be communicated back to stakeholders who have undergone sequencing. The taxonomy is useful in anticipating where in a particular case the uncertainties may be greatest and where they may resolve over time, or not. The aim of the taxonomy was to standardize the language used to describe uncertainties, to differentiate sources of uncertainty that are malleable, to inform clinical interactions, and to standardize how research questions that address uncertainties are framed. It should be considered dynamic so as to develop in conjunction with translational genomics.

PERSONAL UNCERTAINTIES IN GENOME SEQUENCING

Personal uncertainties, as defined in the taxonomy, involve those related to the psychosocial and existential issues that are important to the individual, including the personal meaning of illness or health information. Personal uncertainties may be strong determinants of reactions and behaviors by patients in response to uncertain information. Several qualitative studies have described personal uncertainties related to uncertain genetic test results. In a qualitative analysis by Solomon (2013), 20 individuals identified to have a VUS in a Lynch syndrome gene described personal uncertainties related to whether their result meant they had Lynch syndrome, and the meaning of the result for their future cancer risks and their family members' cancer risks. Solomon also identified affective responses to receiving a VUS result including disappointment, frustration, sadness, and relief. These results are similar to a qualitative analysis of findings from focus groups by Biesecker et al. (2014) that explored perception of uncertainties related to genome sequencing. In this study, participants described personal uncertainties related to information they may receive, and whether to act on uncertain results, resulting in a discussion of whether the results and the researchers can be fully trusted (Biesecker et al., 2014). Responses from this study led to the development of a reliable and valid scale, the Perceptions of Uncertainty in Genomic Sequencing (PUGS) scale aimed at assessing personal uncertainties of genome sequencing in three domains; clinical, affective, and evaluative (Biesecker et al., 2017). Data from this scale can be used to amass understanding of the areas of greatest practical uncertainties for research participants and suggest areas to be addressed by genetic counselors or other healthcare providers to help form accurate perceptions of uncertain health risk information, how to act on it, and whether to trust it. Use of the PUGS across diverse patient samples can also facilitate meta-analysis of perceived uncertainties to achieve generalizable evidence. Further studies relating perceptions of personal utilities to health outcomes will help to assess its importance as a predictor of affective and behavioral outcomes of genome sequencing.

THE CASE OF VARIANTS OF UNCERTAIN SIGNIFICANCE

Vos, Jansen, et al. (2011) demonstrated that patients interpreted genomic results more accurately when the result, inheritance, and cancer risks were communicated in multiple formats, and if a flyer was provided that explained genetic counseling practice. Patients who were told about the possibility of a VUS result during the first meeting with a genetic counselor, also recalled and interpreted the test results more accurately. Finally, patients interpreted genomic test results better and perceived fewer uncertainties when the result was delivered in an in-person meeting with a genetic counselor, and when the counselor gave explicit attention to patients' affect during the sessions. These findings may be used by genetic counselors to enhance the accuracy of recall and to help counselees cope with the uncertainty to make well-informed decisions about their medical management.

Also in the cancer context, a study of 603 genetic variants identified through clinical genetic testing for inherited cancer susceptibility, identified a substantial proportion (37%) that were classified as a variant of uncertain clinical significance (VUS). The VUS result presents challenges to patients who were hoping for clarity about their risk status but instead learn that no further information beyond their medical and family history is available. However, studies investigating those with a VUS show significant variation in patient interpretation with some perceiving the result as benign and some as pathogenic (Richter et al., 2013; Solomon, 2013; Vos et al., 2008). In studies specific to VUS results, risk perception has typically been operationalized as perception of cancer risk. A study of 36 VUS carriers found that 93% of participants perceived a change in their cancer risk after disclosure of a VUS result, of which 32% perceived their risk for breast cancer as higher post disclosure (Richter et al., 2013). However, a prior study by van Dijk et al. (2004) found that perceptions of breast cancer risk did not change significantly post-disclosure among VUS carriers. Perceptions of the meaning and significance of uncertain genetic testing results on health risk are influenced by cognitions, affect, and motivating factors, as described in Chapter 4, Judgment and Decision-Making in Genome Sequencing. These factors frequently lead to distortions in perceived risk independently of high levels of comprehension post-disclosure (van Dijk et al., 2004). Inconsistencies in VUS interpretation are due to human tendencies to imbue risk with subjective meaning.

THE POTENTIAL HEALTH THREAT OF UNCERTAINTIES

The psychological literature offers several theoretical models about how patients cope with health-threatening information. Two models that have recently gained attention are Park and Folkman's Transactional Model of Stress and Coping (TMSC) and Mishel's Perceived Uncertainty in Illness Theory (Mishel, 1988, 1990; Park & Folkman, 1997). Mishel's Perceived Uncertainty in Illness Theory is similar to the TMSC in that it posits that responses to receipt of uncertain health information can lead to perceptions of a health threat and appraised for the type and degree of threat. Mishel observed that uncertainties have the potential to be perceived as danger or opportunity. In either case, these appraisals can lead to perceptions that mitigate negative responses, such as notions of personal control.

The TMSC models the dynamic process of making meaning of a health threat that is influenced by personal and environmental characteristics, and the appraisals made of the threat (Park & Folkman, 1997). It posits that when faced with the stress of receiving an uncertain genetic test result, an individual evaluates the personal significance of what is happening (primary appraisal) and what can be done about the result (secondary appraisal). As circumstances and an individual's values, beliefs, and goals (global meaning) change over time, so will their appraisal, coping, and adaptation to the health threat. Appraisals and means of coping are dynamic and influenced by personality traits and changes in perceptions of the health threat. Similarly, the theory of uncertainty management, developed by Brashers, Neidig, & Goldsmith (2004), focuses on peoples' responses to communication of uncertain health information from a healthcare provider. The theory posits four strategies for managing perceptions of uncertainty over time: information seeking, adapting to chronic uncertainty, relying on social support and personal resources to manage uncertainty, and managing a need for certainty in one area with a desire for uncertainty in another. Strategies are taken according to circumstances, past experiences, and personality traits (Carcioppolo, Yang, & Yang, 2016). The next sections will focus on these individual differences, and specific strategies used by counselees to cope with threatening information from genomic testing.

INDIVIDUAL DIFFERENCES IN TOLERANCE OF UNCERTAINTIES

Confrontation with the uncertain reality of genetic testing may lead to disappointment and uncertainty among those who characteristically avoid ambiguity or have a lower tolerance of uncertainty (Han et al., 2009; Hillen et al., 2017). Patients' perceived need for certainty may collide with their perceived certainty of the test results or risk as outcomes of genetic counseling. It is not merely the perceived uncertainties of the current situation that influence the wellbeing and decision-making of patients, but also the degree of unwanted certainty. If a patient does not have a strong desire to receive certain information when they come to genetic counseling, they will be more satisfied, and better able to cope with an uncertain test result than a patient with a strong need for certainty. This was demonstrated in a study in 467 patients who had received a genetic test result for hereditary breast and ovarian cancer (Vos, Menko, et al., 2013). Vos and colleagues found that before and after receiving test results, most women experienced an unfulfilled need for certainty about the test results, the role of heredity, and their cancer risk. The return of a pathogenic mutation decreased uncertainty about the test results but increased uncertainty about cancer and family inheritance. The communication of a VUS result did not fulfill the women's needs for certainty. In all domains of uncertainty, women with a VUS test result experienced less fulfillment of their need for certainty than those with a pathogenic mutation or a benign variant result. The women differentiated the unfulfilled need for certainty between the domains of cancer, genetic test results, and inheritance. The unfulfilled need for certainty was not correlated with the women's cognitive understanding of their genetic test results.

Vos, Stiggelbout, et al. (2011) identified differences among individuals in the extent they preferred to learn certainty about their health. Some individuals have greater tolerance of uncertainty in a specific domain of their life than others. This trait has been described in the literature as tolerance of ambiguity as well. Hillen et al. (2017) reviewed eight scales developed to assess this trait, calling attention to the different ways it has been defined and assessed. They compared tolerance of ambiguity to ambiguity aversion; pessimistic appraisals of ambiguous risks and choice options, and the avoidance of decision-making. Similarly, ambiguity aversion has been shown to have health consequences. For example, individuals who perceive greater ambiguity in health information perceive themselves to be at higher risk for illness, view illness as less preventable, and engage in fewer preventive behaviors (Han, Moser, & Klein, 2007). Taber et al. (2015) reported that participants who perceived sequencing results as more ambiguous were inclined to have less favorable cognitions about results and lower intentions to learn and share results with relatives. Among those who had low tolerance for uncertainty or were less optimistic, greater perceived ambiguity was associated with lower intentions to learn results for nonmedically actionable diseases.

Taber's results are consistent with the phenomenon of ambiguity aversion that may influence whether people learn and communicate genomic information.

INTERPRETATION OF UNCERTAIN GENOMIC INFORMATION

Most individuals receiving uncertain results differentiate between their recollection and interpretation of genetic information communicated by their genetic healthcare provider (Cypowyj et al., 2009; Solomon, 2013; Vos et al., 2008). This means that they identify a difference between what was communicated by the healthcare provider and how they subjectively interpret the result. In a qualitative study of 24 women with a VUS in *BRCA1/2* genes, most women in the study had a different interpretation than their recollection, with 79% interpreting their VUS result as a pathogenic variant and 21% as a true-negative result (Vos et al., 2008). The investigators proposed that when faced with uncertain genetic results individuals interpret the result with greater certainty as an adaptive way of coping with uncertain information by minimizing the uncertainty (see Chapter 4, Judgment and Decision-Making in Genome Sequencing, for a discussion of factors affecting risk perceptions). According to this hypothesis, perceptions of uncertainty and appraisals may influence a client's reinterpretation of genetic information communicated by a healthcare provider.

Responses in the qualitative study led to the development of a valid and reliable scale aimed at assessing an individual's recollections and interpretations of both their cancer risks and hereditary likelihood (Vos, Oosterwijk, et al., 2011). This scale was used retrospectively in a study of 206 patients who had undergone *BRCA1/2* testing in which 76 received a VUS result, 76 a benign result, and 51 a pathogenic variant. Information was gathered from patient visit summary letters to compare healthcare provider recorded communication of genetic information to the patients' recollections and interpretations. Differences in recollections and interpretations of cancer risks and hereditary likelihood were found to be greatest among those who received a VUS result. Furthermore, the majority with a VUS overestimated their cancer risks and hereditary likelihood. These findings support the hypothesis that patients reinterpret genetic information as a form of meaning-based coping. Furthermore, the observation that greater differences between recollection and interpretations were observed in the VUS group may be explained only in part by the uncertainty of the result and warrants further investigation. Among the 76 participants in the VUS group; psychological outcomes, quality of life, *BRCA 1/2*-related stigma and vulnerability, and medical outcomes were better predicted by the participants' interpreted cancer risks and hereditary likelihoods than by the communicated information (Vos, van Asperen, et al., 2013). Those who interpreted their risks as greater than what was communicated by their genetic healthcare provider had more worries, psychological distress, and greater uptake in prophylactic surgeries. Collectively, these findings suggest that individuals receiving a VUS may appraise the result more often as a danger leading to higher perceptions of cancer risks and inherited likelihood, and surgical risk management. Yet, there is scant evidence among systematic reviews that overestimation of cancer risks leads to excessive surveillance or prophylactic surgery due to inflated risk perception (Meiser et al., 2006–2007; Yanes, Willis, Meiser, Tucker, & Best, 2018). Similarly, underestimation does not lead to excessive avoidance of medical treatment options such as regular surveillance (Meiser et al., 2006–2007; Yanes et al., 2018).

Borrowing from the dual-process theory of Kahneman and Tversky, researchers have made a distinction between cognitive and affective factors that lead to distinct risk perceptions (see Chapter 4: Judgment and Decision-Making in Genome Sequencing). Studies from Vos, Jansen, et al. (2011) and Bonner (2017) demonstrate that the interpretations of genomic results include thoughts and feelings. Vos and Bonner found that individuals provided different answers when they are asked about their thoughts ("what do you think is your likelihood of developing cancer?") and feelings ("how do you feel about your likelihood of developing cancer?"). It seems that the question about knowledge may have focused more on the recollection of what the genetic counselor had communicated, and the question about feelings on their subjective interpretation. Bonner (2017) expanded on these findings by assessing recollection of results, thoughts about VUS results, and feelings about VUS results, among breast cancer survivors and found there to be three distinct risk perceptions, suggesting that cognitive risk is not simply a recollection of risk. While past studies have suggested that affective risk perceptions (feeling judgments) are the strongest predictor of health outcomes, the TRIRISK model demonstrates that each component of risk perception, cognitive, affective, and experiential, plays a novel role in predicting specific health outcomes (Ferrer, Klein, Persoskie, Avishai-Yitshak, & Sheeran, 2016). Taken together, these studies represent a rich and dynamic literature on risk perceptions. More evidence is needed to understand the many interrelated factors and traits that influence perceptions of uncertainties, their relation to risk perceptions, and how they predict health outcomes. Genome sequencing offers the opportunity for further research in this area.

HEALTHCARE CONSEQUENCES OF UNCERTAINTIES

Patients reinterpret communicated test results, likely as a way of coping with the unwanted uncertainty of the genomic test result. This reinterpretation also influences their healthcare decisions. Several studies have reported frequencies of prophylactic surgeries after receiving a VUS result that overlap with rates in cohorts of patients with a pathogenic variant, raising concern that patients and providers may be attributing meaning to an uncertain genetic test result (Murray, Cerrato, Bennett, & Jarvik, 2011; Vos et al., 2008). A retrospective chart review of 107 VUS carriers found that

13 of 22 individuals who had risk-reducing oophorectomy reported that uncertainty related to their genetic test result influenced their decision (Murray et al., 2011). Similarly, a prospective study of 183 women with a benign result showed similar intentions for mammogram screening as those who received a pathogenic variant result (van Dijk et al., 2005). These data suggest clients' reactions to residual uncertainty related to their genetic result influence how they make decisions about their future healthcare management. Patients with a strong cancer history may be more likely to imbue uncertain results with greater potential for pathogenicity. Understanding how clients perceive residual uncertainties related to their result and how these perceptions influence their healthcare decisions has implications for how uncertainty is most effectively communicated and healthcare options assessed with clients in posttest counseling discussions.

Few studies of return of genomic test results have addressed outcomes among pregnant couples receiving uncertain information about their unborn fetus. Sapp et al. (2010) reported on results from interviews with 36 women who expressed ambivalence about decisions to undergo amniocentesis. While all the women interviewed had undergone prenatal testing based on reported feelings of parental responsibility, the majority expressed significant ambivalence about their decision, relaying concerns that they may have taken undue risks for the reassurance of a normal result (Sapp et al., 2010). Tensions were evident among intellectual, moral, and spiritual values that related to making a decision to undergo amniocentesis. Even after receiving negative results, perceptions of ambivalence about the decision remained. Perhaps if the women had deliberated some of the competing tensions during genetic counseling while making their testing decision, some of the ambivalence may have been resolved to enhance the quality of the women's decisions.

In the case of receipt of unknown prenatal results, fetal microarray results encompass significant uncertainties. In 2%−3% of women undergoing prenatal screening, microarray testing detects deletions and duplications in a fetus' genome that go undetected by conventional means. Many of these copy number variants are associated with variable or uncertain phenotypes in children. Bernhardt et al. (2013) interviewed 23 pregnant women who received uncertain fetal microarray results. Most described learning the uncertain finding to be shocking and overwhelming (Bernhardt et al., 2013). As the women worked to learn more about the results, they experienced lingering uncertainties. Some women felt that by working hard at it, they could eventually uncover more definitive information about their fetus to make an informed decision about the pregnancy. Others described receiving and reading contradictory information about the results and expressed confusion about how to proceed. When the result offered an explanation of why a fetus had a physical abnormality detected by ultrasound, the result was interpreted as a confirmation of a condition that facilitated decisions to terminate the pregnancy.

Werner-Lin, McCoyd, and Bernhardt (2016) interviewed 24 members of 12 couples who received uncertain fetal microarray results. These parents struggled with the unknown nature of the information, expressing frustration that they were unprepared to receive such results and that clarifying information was not available (Werner-Lin et al., 2016). Male partners more often dismissed their own distress about the results in an effort to support their partners. These couples had no precedent for how to manage these uncertainties and thus chose language to describe their circumstances that merged their experiences with those of other parents who face uncertainty about their future child.

Following birth of the children of these couples, Werner-Lin, Walser, Barg, and Bernhardt (2017) conducted follow-up interviews with the parents who largely reported their infants to be developing normally. The parents expressed concern about their child's future development but described ways they used to reassure themselves that their child was unaffected (Werner-Lin et al., 2017). They reported comparing their infant to his/her siblings, monitoring the baby's development and behaviors, and taking to heart reassurance they heard from healthcare providers. Some parents reported being acutely aware of their infants' neurocognitive development and stated that if atypical development were to arise, they would attribute it to the uncertain copy number variant results.

These prenatal studies are small and exploratory, but suggest potential negative consequences of unexpected uncertain information about one's developing fetus. Resources to help parents expect the possibility of uncertain results and strategies to help them manage the surprise and lack of resources would be advantageous. Outcomes of prior studies described in this chapter suggest that parents are prone to unconsciously distorting the risks to their children, as they imbue the uncertainties with subjective meaning.

ESTABLISHING EXPECTATIONS FOR UNCERTAINTIES

Turbitt et al. (2018) reported on results from a randomized controlled trial comparing standard consent to genome sequencing to a shorter, lower literacy consent process that found equivalence in the outcomes. Yet in a secondary analysis only 60% of the cohort of women with premature ovarian insufficiency who consented to genome sequencing, made an informed decision to learn secondary findings. The participants were focused on their primary diagnosis as the reason they were undergoing sequencing. While most opted to receive secondary findings, they were not all clear that these were medically actionable results unrelated to their primary condition and that some would come with uncertainties. When consenting patients to receive secondary findings, clear differentiation from the primary indication and the likelihood of receiving uncertain results should be deliberated.

Few studies have been done on the communication of uncertainties in genomic sequencing, either at the time of consenting participants, or when delivering VUS results. As such, there are key research gaps about how this information is most effectively communicated to patients and research participants and what may be the most effective ways to describe VUS results so as not to alarm patients yet help them manage them over time. In the face of uncertain risk, it is sensible to frame the unknown positively as opportunity for taking control by acting on the risk and for the hope offered by reducing risk. An uncertain result does not diminish the opportunity for screening and monitoring unclear risk for many conditions. Furthermore, framing the uncertain risk information as positive for sharing with family members can lead to opportunities for others to take action if they are concerned about a residual familial risk.

CONCLUDING REMARKS

Much has been made of the large amount of information generated from genomic analysis and initial ideas on how to provide this information to potential stakeholders. It is customary in the tradition of medicine, and specifically in clinical genetics, to inform clients thoroughly so that they can make an informed decision (about testing or otherwise) that is in line with their personality, beliefs, and moral values. We have 30 years of experience in which the information was mainly about rare diseases caused by a single-gene mutation. In a number of cases this mutation could result in different phenotypical manifestations, which made the information more complex. Genetic counseling should entail communication of all relevant information, as it may be more important to process the most relevant information to the patient to make an informed choice. In the case of single-gene conditions this has been shown to be an achievable, although complicated, undertaking. With genome sequencing, maximum information provision is not possible. This begs the question of what we should strive to convey, because the eventual decision is a matter of learning information that will challenge one's ability to also deal with uncertainty. Research has shown that people are inclined to give meaning to results of genetic testing, uncertain or not, and that the objective information they have received is often of secondary importance.

The lessons from single-gene testing show that individuals generally experience minimal adverse psychological consequences after learning test results (Broadstock, Michie, & Marteau, 2000; Paulsen et al., 2013). Despite the complexities of genomic sequencing, patients who receive uncertain genomic information are also largely resilient in the face of uncertainties (Gray et al., 2014). However, though anxiety and depressive symptoms may not exceed clinically relevant levels, people differ in experiencing complexities and problems, not least because of personality characteristics, moral values, and support systems. We live in an uncertain world that most people have the capacity to navigate. And while we generally tend toward understanding the world around us and how it works in our lives, we also come to accept that there are many dimensions that we do not fully understand or can predict. In genetic counseling, patients with high optimism and low depressive symptoms perceive less uncertainty than those without these characteristics and have fewer negative consequences. Clients can be helped to expect uncertain results, and to consider ways that they may be biased to over- or under-interpret their meaning. Spending time comparing VUS results to other types of results may help clients to have more realistic risk perceptions. Overall, clients can be helped to frame uncertainty not only as opportunity and hope for an unknown future, but also as an avenue for personal growth.

REFERENCES

Bernhardt, B. A., Soucier, D., Hanson, K., Savage, M. S., Jackson, L., & Wapner, R. J. (2013). Women's experiences receiving abnormal prenatal chromosomal microarray testing results. *Genetics in Medicine*, 15(2), 139–145. Available from https://doi.org/10.1038/gim.2012.113.

Biesecker, B. B., Klein, W., Lewis, K. L., Fisher, T. C., Wright, M. F., Biesecker, L. G., & Han, P. K. (2014). How do research participants perceive "uncertainty" in genome sequencing? *Genetics in Medicine*, 16(12), 977–980. Available from https://doi.org/10.1038/gim.2014.57.

Biesecker, B. B., Woolford, S. W., Klein, W. M. P., Brothers, K. B., Umstead, K. L., Lewis, K. L., & Han, P. K. J. (2017). PUGS: A novel scale to assess perceptions of uncertainties in genome sequencing. *Clinical Genetics*, 92(2), 172–179. Available from https://doi.org/10.1111/cge.12949.

Bloss, C. S., Zeeland, A. A., Topol, S. E., Darst, B. F., Boeldt, D. L., Erikson, G. A., ... Torkamani, A. (2015). A genome sequencing program for novel undiagnosed diseases. *Genetics in Medicine*, 17(12), 995–1001. Available from https://doi.org/10.1038/gim.2015.21.

Bonner, D. (2017). Understanding how clients make meaning of their variants of uncertain significance in the age of multi-gene panel testing for cancer susceptibility (MSc). John Hopkins University, Baltimore, MD.

Brashers, D. E., Neidig, J. L., & Goldsmith, D. J. (2004). Social support and the management of uncertainty for people living with HIV or AIDS. *Health Communication*, 16(3), 305–331.

Broadstock, M., Michie, S., & Marteau, T. (2000). Psychological consequences of predictive genetic testing: A systematic review. *European Journal of Human Genetics*, 8(10), 731–738. Available from https://doi.org/10.1038/sj.ejhg.5200532.

Carcioppolo, N., Yang, F., & Yang, Q. (2016). Reducing, maintaining, or escalating uncertainty? The development and validation of four uncertainty preference scales related to cancer information seeking and avoidance. *Journal of Health Communication*, 21(9), 979–988. Available from https://doi.org/10.1080/10810730.2016.1184357.

Cypowyj, C., Eisinger, F., Huiart, L., Sobol, H., Morin, M., & Julian-Reynier, C. (2009). Subjective interpretation of inconclusive BRCA1/2 cancer genetic test results and transmission of information to the relatives. *Psychooncology*, 18(2), 209–215. Available from https://doi.org/10.1002/pon.1407.

Ferrer, R. A., Klein, W. M., Persoskie, A., Avishai-Yitshak, A., & Sheeran, P. (2016). The tripartite model of risk perception (TRIRISK): Distinguishing deliberative, affective, and experiential components of perceived risk. *Annals of Behavioral Medicine*, 50(5), 653–663.

Gray, S. W., Martins, Y., Feuerman, L. Z., Bernhardt, B. A., Biesecker, B. B., Christensen, K. D., & members of the CSER Consortium Outcomes and Measures Working Group. (2014). Social and behavioral research in genomic sequencing: Approaches from the Clinical Sequencing Exploratory Research Consortium Outcomes and Measures Working Group. *Genetics in Medicine*, 16(10), 727–735. Available from https://doi.org/10.1038/gim.2014.26.

Han, P. K., Klein, W. M., & Arora, N. K. (2011). Varieties of uncertainty in health care: A conceptual taxonomy. *Medical Decision Making*, *31*(6), 828−838.

Han, P. K., Moser, R. P., & Klein, W. M. (2007). Perceived ambiguity about cancer prevention recommendations: Associations with cancer-related perceptions and behaviours in a US population survey. *Health Expectations*, *10*(4), 321−336. Available from https://doi.org/10.1111/j.1369-7625.2007.00456.

Han, P. K., Reeve, B. B., Moser, R. P., et al. (2009). Aversion to ambiguity regarding medical tests and treatments: Measurement, prevalence, and relationship to sociodemographic factors. *Journal of Health Communication*, *14*, 556−572.

Harper, P. S. (2014). Huntington's Disease in a historical context. In G. P. Bates, S. J. Tabrizi, & L. Jones (Eds.), *Huntington's disease* (4th ed., pp. 3−24). New York: Oxford University Press.

Hillen, M. A., Gutheil, C. M., Strout, T. D., Smets, E. M., & Han, P. K. (2017). Tolerance of uncertainty: Conceptual analysis, integrative model, and implications for healthcare. *Social Science & Medicine*, *180*, 62−75. Available from https://doi.org/10.1016/j.socscimed.2017.03.024.

Meiser, B., Gaff, C., Julian-Reynier, C., Biesecker, B. B., Esplen, M. J., Vodermaier, A., & Tibben, A. (2006). International perspectives on genetic counseling and testing for breast cancer risk. *Breast Disease*, *27*, 109−125.

Mishel, M. H. (1988). Uncertainty in illness. *Image—The Journal of Nursing Scholarship*, *20*(4), 225−232.

Mishel, M. H. (1990). Reconceptualization of the uncertainty in illness theory. *Image—The Journal of Nursing Scholarship*, *22*(4), 256−262.

Murray, M. L., Cerrato, F., Bennett, R. L., & Jarvik, G. P. (2011). Follow-up of carriers of BRCA1 and BRCA2 variants of unknown significance: Variant reclassification and surgical decisions. *Genetics in Medicine*, *13*(12), 998−1005. Available from https://doi.org/10.1097/GIM.0b013e318226fc15.

Park, C. L., & Folkman, S. (1997). Meaning in the context of stress and coping. *Review of General Psychology*, *1*(2), 115−144.

Paulsen, J. S., Nance, M., Kim, J. I., Carlozzi, N. E., Panegyres, P. K., Erwin, C., ... Williams, J. K. (2013). A review of quality of life after predictive testing for and earlier identification of neurodegenerative diseases. *Progress in Neurobiology*, *110*, 2−28. Available from https://doi.org/10.1016/j.pneurobio.2013.08.003.

Richter, S., Haroun, I., Graham, T. C., Eisen, A., Kiss, A., & Warner, E. (2013). Variants of unknown significance in BRCA testing: Impact on risk perception, worry, prevention and counseling. *Annals of Oncology*, *24*(Suppl 8), viii69−viii74. Available from https://doi.org/10.1093/annonc/mdt312.

Sapp, J. C., Hull, S. C., Duffer, S., Zornetzer, S., Sutton, E., Marteau, T. M., & Biesecker, B. B. (2010). Ambivalence toward undergoing invasive prenatal testing: An exploration of its origins. *Prenatal Diagnostics*, *30*(1), 77−82. Available from https://doi.org/10.1002/pd.2343.

Solomon, I. (2013). Living in Lynch syndrome limbo: Understanding the meaning of uncertain genetic test results (Unpublished master's thesis, MSc). John Hopkins University, Baltimore, MD.

Taber, J. M., Klein, W. M., Ferrer, R. A., Han, P. K., Lewis, K. L., Biesecker, L. G., & Biesecker, B. B. (2015). Perceived ambiguity as a barrier to intentions to learn genome sequencing results. *Journal of Behavioral Medicine*, *38*(5), 715−726. Available from https://doi.org/10.1007/s10865-015-9642-5.

Turbitt, E., Roberts, M. C., Ferrer, R. A., Taber, J. M., Lewis, K. L., Biesecker, L. G., ... Klein, W. M. (2018). Intentions to share exome sequencing results with family members: Exploring spousal beliefs and attitudes. *European Journal of Human Genetics*, *26*(5), 735−739. Available from https://doi.org/10.1038/s41431-018-0118-2.

van Dijk, S., Otten, W., Timmermans, D. R., van Asperen, C. J., Meijers-Heijboer, H., Tibben, A., ... Kievit, J. (2005). What's the message? Interpretation of an uninformative BRCA1/2 test result for women at risk of familial breast cancer. *Genetics in Medicine*, *7*(4), 239−245. https://doi.org/1010.109701.GIM.0000159902.34833.26.

van Dijk, S., van Asperen, C. J., Jacobi, C. E., Vink, G. R., Tibben, A., Breuning, M. H., & Otten, W. (2004). Variants of uncertain clinical significance as a result of BRCA1/2 testing: Impact of an ambiguous breast cancer risk message. *Genetic Testing*, *8*(3), 235−239. Available from https://doi.org/10.1089/gte.2004.8.235.

Vos, J., Jansen, A. M., Menko, F., van Asperen, C. J., Stiggelbout, A. M., & Tibben, A. (2011). Family communication matters: The impact of telling relatives about unclassified variants and uninformative

DNA-test results. *Genetics in Medicine*, *13*(4), 333−341. Available from https://doi.org/10.1097/GIM.0b013e318204cfed.

Vos, J., Menko, F. H., Oosterwijk, J. C., van Asperen, C. J., Stiggelbout, A. M., & Tibben, A. (2013). Genetic counseling does not fulfill the counselees' need for certainty in hereditary breast/ovarian cancer families: An explorative assessment. *Psychooncology*, *22*(5), 1167−1176. Available from https://doi.org/10.1002/pon.3125.

Vos, J., Oosterwijk, J. C., Gomez-Garcia, E., Menko, F. H., Jansen, A. M., Stoel, R. D., . . . Stiggelbout, A. M. (2011). Perceiving cancer-risks and heredity-likelihood in genetic-counseling: How counselees recall and interpret BRCA 1/2-test results. *Clinical Genetics*, *79*(3), 207−218. Available from https://doi.org/10.1111/j.1399-0004.2010.01581.x.

Vos, J., Otten, W., van Asperen, C., Jansen, A., Menko, F., & Tibben, A. (2008). The counsellees' view of an unclassified variant in BRCA1/2: Recall, interpretation, and impact on life. *Psychooncology*, *17*(8), 822−830. Available from https://doi.org/10.1002/pon.1311.

Vos, J., Stiggelbout, A. M., Oosterwijk, J., Gomez-Garcia, E., Menko, F., Collee, J. M., . . . Tibben, A. (2011). A counselee-oriented perspective on risk communication in genetic counseling: Explaining the inaccuracy of the counselees' risk perception shortly after BRCA1/2 test result disclosure. *Genetics in Medicine*, *13*(9), 800−811. Available from https://doi.org/10.1097/GIM.0b013e31821a36f9.

Vos, J., van Asperen, C. J., Oosterwijk, J. C., Menko, F. H., Collee, M. J., Gomez Garcia, E., & Tibben, A. (2013). The counselees' self-reported request for psychological help in genetic counseling for hereditary breast/ovarian cancer: Not only psychopathology matters. *Psychooncology*, *22*(4), 902−910. Available from https://doi.org/10.1002/pon.3081.

Werner-Lin, A., Barg, F. K., Kellom, K. S., Stumm, K. J., Pilchman, L., Tomlinson, A. N., & Bernhardt, B. A. (2016). Couple's narratives of communion and isolation following abnormal prenatal microarray testing results. *Qualitative Health Research*, *26*(14), 1975−1987.

Werner-Lin, A., McCoyd, J. L., & Bernhardt, B. A. (2016). Balancing genetics (science) and counseling (art) in prenatal chromosomal microarray testing. *Journal of Genetic Counseling*, *25*(5), 855−867. Available from https://doi.org/10.1007/s10897-016-9966-5.

Werner-Lin, A., Walser, S., Barg, F. K., & Bernhardt, B. A. (2017). "They can't find anything wrong with him, yet": Mothers' experiences of parenting an infant with a prenatally diagnosed copy number variant (CNV). *American Journal of Medical Genetics Part A*, *173*(2), 444−451. Available from https://doi.org/10.1002/ajmg.a.38042.

Yanes, T., Willis, A. M., Meiser, B., Tucker, K. M., & Best, M. (2018). Psychosocial and behavioral outcomes of genomic testing in cancer: A systematic review. *European Journal of Human Genetics*. https://doi.org/10.1038/s41431-018-0257-5. [Epub ahead of print].

FURTHER READING

Kaphingst, K. A., Ivanovich, J., Biesecker, B. B., Dresser, R., Seo, J., Dressler, L. G., . . . Goodman, M. S. (2016). Preferences for return of incidental findings from genome sequencing among women diagnosed with breast cancer at a young age. *Clinical Genetics*, *89*(3), 378−384. Available from https://doi.org/10.1111/cge.12597.

CHAPTER 6

Direct-to-Consumer Genetic Testing

A. Cecile J.W. Janssens
Department of Epidemiology, Rollins School of Public Health, Emory University, Atlanta, GA, United States

At the turn of the millennium, when the Human Genome Project was about to publish the first full sequence of human DNA, about 100 websites were offering direct-to-consumer (DTC) genetic testing (Gollust, Wilfond, & Hull, 2003). Most sites offered DNA testing to determine paternity, but about 15% offered health-related genetic tests that determined carrier status of DNA mutations for hereditary disorders or recommended nutrients for a healthier lifestyle. It was around the same time that Myriad Genetic Laboratories started advertising testing *BRCA* mutations for breast and ovarian cancer on TV and radio and in newspapers and magazines (Ramos & Weissman, 2018). While their tests still needed to be ordered by physicians, it was the first campaign to raise awareness about the opportunities for DNA testing in the general population.

About five years later, in 2007, when it had become affordable to genotype 100,000s of DNA markers on a single chip and genome-wide association studies (GWAS) were reporting their first discoveries, 23andMe was the first to offer a broad personal genome service directly to consumers. This service included polygenic prediction of common diseases, carrier testing, ancestry, and fun-to-know characteristics like earwax type, alcohol flush reaction, and muscle performance. Several companies followed, including deCODEme, the commercial spin-off of the pioneering and prolific deCODE research institute.

Meanwhile, the US Food and Drug Administration (FDA) closely monitored the developments in this new market. The agency did not yet know how to regulate these new tests, but it knew what it did not want. In 2010, the FDA sent letters to a dozen companies, warning that they were offering medical tests, including some for drug interactions, for which they had not requested premarket approval. All companies were requested to provide the necessary information about the analytical and clinical validity of their tests. Unsurprisingly, the companies could not answer these questions and closed their businesses in subsequent years: from a scientific point of view, their tests were not expected to have clinical validity, and the companies did not have the relevant data to prove otherwise.

In 2013, the FDA again urged 23andMe to stop the marketing of its personal genome service because the agency had not cleared or approved the marketing of the

Clinical Genome Sequencing
DOI: https://doi.org/10.1016/B978-0-12-813335-4.00006-4

BRCA test and the pharmacogenetic tests (Annas & Elias, 2014). 23andMe had not provided the data that the agency had requested. The company could not provide it, nor could scientists: the data about clinical validity did not exist (Janssens, 2014). For two years, the company only provided ancestry testing to the public and evolved as a major research company while it worked with the FDA on permission to resume the offer of health-related tests. The company was authorized to offer genetic testing for Bloom's disease and other recessive disorders in 2015 and to offer genetic risk predictions in 2017.

It is worth emphasizing that the FDA did not approve these tests, but authorized their offer. Authorization means that the agency considers these tests are low- to moderate-risk devices that do not need FDA approval. The agency put "special controls" in place that should ensure that people receive proper and detailed information, before and after testing, about the accuracy of the test, its predictive ability, strengths, and limitations, as well as resources for further information. FDA authorization does not mean that the agency finds that the tests have clinical utility or that they can meaningfully inform about the risk of disease. They leave that up to the consumer to decide.

More recently, in 2017, at a time when genome sequencing became increasingly affordable, Illumina, the market leader in sequencing technology, launched the Helix platform, an online marketplace for companies that offer genetic tests. Illumina sequences the DNA for all customers of these companies, who can, in turn, offer their services at a discounted rate to customers when they have already been genotyped.

Commercial DNA testing is booming, but this brief historical overview shows that it has already done so twice before. Time magazine considered the personal genome service of 23andMe the number one invention of the year in 2008, but the market imploded a few years later. Time will tell whether the business is here to stay—it may be here to stay for some tests, but not all.

THE ONLINE OFFER OF TESTS

What DTC genetic tests have in common is that customers order them at their initiatives, not through a healthcare professional, and that may be the only shared characteristic. The DTC offer of genetic tests varies so widely that it is safe to say that no two offers are the same. For understanding the psychosocial and behavioral consequences of DTC genetic testing, it is therefore essential to start with the questions: what disease or phenotype is tested, how is the test offered, and why is the test ordered?

Health-related DTC genetic tests are offered for a wide range of diseases and risk factors. They generally fall in the following categories:
— Carrier testing, which tests the presence or absence of recessive and dominant mutations for hereditary diseases and syndromes;

- Pharmacogenetic testing, which assesses genetic markers that impact the efficacy and adverse effects of drug treatment;
- Polygenic risk scores, which inform about the susceptibility to disease; and
- Wellness genomics, which reports DNA variations that are used to personalize nutritional and lifestyle recommendations.

The online offer varies substantially in *how* a test is offered. A genetic test may be offered separately or as part of a broader panel of tests. Some companies offer DNA tests in one category, for example, only carrier testing to prospective parents, genetic testing for hereditary diseases via physicians, or nutrigenomic and lifestyle tests, while others offer all or most of the above in broad personal genome services that often also include ancestry testing. Some test for a limited number of diseases, others test hundreds.

The offer of DTC genetic tests also varies in the amount of information and the quality of the service that is provided. Some companies have a simple overview of the tests they offer, with minimal description to explain, while others offer detailed information about the test, the disease, opportunities for prevention, and so on. Some companies go the extra mile to explain what DNA is, how it impacts health, and what customers can do with the genetic information. This information may include details about the test, the presentation of the test results, and the provision of educational materials, guidance for next steps, and genetic counseling (Lachance, Erby, Ford, Allen, & Kaphingst, 2010). Companies may also offer access to genetic counseling.

When the offer of genetic tests varies so widely, it is important to keep in mind that customers may have received their carrier status or risks for a specific disease for other reasons than that they were interested to know. People may buy broad personal genome services because they want to know their genetic predisposition for one disease, and may not know what other diseases they will be informed about until they see the results. Other people may have a general interest in health, or may be interested in learning about their ancestry and received complimentary health-related genetic test results. Customers who received genetic test results for diseases that they were not expecting may be more unprepared to cope with the results.

CRITICAL APPRAISAL OF SCIENTIFIC STUDIES ON DTC GENETIC TESTING

The rapid discoveries in genomic research, largely delivered by GWAS, lead to the ongoing development and continuous improvement of commercial genetic tests and test panels. The online offer has rapidly changed and expanded over time and will continue to do so. As a result, almost every scientific study on DTC genetic testing has investigated a different test or service.

When a study concludes that people have a reasonable understanding of the test (Gordon et al., 2012), that they understand that genetic predisposition is not deterministic (Kaphingst et al., 2012), it means that they understood that for the test under study. It does not mean that people will also understand the test when it becomes more complex or when it is offered as part of a panel of tests. It also does not mean that they would have understood the test if they had the disease themselves or when the disease ran in their families. The results of the study may be limited to the test under study.

For interpreting the results of scientific studies, it is also essential to know whether the participants were actual consumers, people who are considering the test, or just people from the general population who have heard about the test but have no further interest: did they buy the test, receive it for free, or are they responding to a hypothetical scenario (Janssens, 2010)? If a test has personal relevance and participants bought it because they are eager to learn about their carrier status or genetic risk, then the test may have more psychological or behavioral impact than when the participants are, say, adults responding to a hypothetical scenario.

The differences between the genetic tests that are studied may be most profound for polygenic testing for common diseases, and that will be the focus of the rest of this chapter. Adding more single nucleotide polymorphisms (SNPs) to a polygenic risk test may improve its predictive ability, which in turn may have a different impact on behavioral or emotional responses. Even when the improvement in predictive risk is minimal, the simple fact that the test includes more SNPs may change the *perceived* predictive ability and impact wellbeing and behavior. The impact of polygenic testing may also differ depending on whether one or multiple diseases are tested.

The differences in genetic tests challenge the comparison between studies and the synthesis of knowledge across studies. The findings of a scientific study must be interpreted understanding that they may be specific for the test that was investigated, and may not be observed when the test offer changes.

POLYGENIC RISK SCORES

A polygenic risk score is a number that expresses the combined contribution of multiple SNPs on the risk of disease. The score sums the number of risk alleles for all SNPs in the score. Polygenic risk scores are *unweighted,* when the risk alleles for all SNPs yield similar increases in disease risk, or *weighted* scores when one or more risk alleles increase risk substantially higher than others. The weights are obtained from GWAS. An unweighted polygenic risk score that is based on, say, 10 SNPs for each of which an individual may have 0, 1, or 2 risk alleles, generates scores that range from 0 to 20. An individual who has more risk alleles for these SNPs has a higher polygenic risk score, and a higher risk of disease.

The value of a polygenic score itself has no intrinsic interpretation as the range of values depends on how many SNPs were included in the score and whether weights were considered. That is why, in scientific studies, polygenic risk scores are often transformed to values that have a mean of 0 and a standard deviation of 1, which makes it at least possible to see whether an individual has a score higher or lower than the average of 0. The polygenic risk scores are transformed into risk models using regression analysis. These risk models are developed and validated in scientific studies in which the SNPs and disease status are known for a large group of people. Nongenetic predictors may be added to the risk model.

Companies that predict risks for common diseases do not have these datasets in which they can make and test their predictions. Instead, they rely on prediction algorithms ("templates") and use data from the scientific literature. To give a simple example, if a company predicts the risk of disease for type 2 diabetes using a single SNP, they might find in the literature that the average risk of type 2 diabetes for people with zero risk alleles on the SNP is 20% and that the risk is doubled per risk allele. The formula then tells that a person who carries two risk alleles has a risk of 80% (20% $\times$ 2 $\times$ 2).

The algorithms for the polygenic risk scores that are used by the companies may be more complex, but the example illustrates how a formula and literature data are combined. For the calculation of polygenic risk, companies need an estimate of the average risk of disease for people, ideally for people of the same age and sex as each customer, and they need the weights for all risk alleles of the SNPs that are considered in the score (Kalf et al., 2014). Risk predictions may differ between companies because they use different algorithms, different SNPs, and different scientific data about average risks and weights (Kalf et al., 2014). The accuracy of the predictions is a major concern.

HOW ACCURATE ARE POLYGENIC RISK SCORES?

The accuracy of polygenic risk scores in commercial DNA tests is a major and valid concern, but what accuracy means may not be so clear: when are risk predictions accurate? Many people will say that predictions are accurate when they correctly tell what will happen. Correct telling of what will happen should only be expected for disease risks that are 0% or 100%. These are the only two predictions that claim that the person will not develop the disease or that they will. For all probabilities, the person might or might not develop the disease. Then, what is accuracy?

In prediction research, accuracy is measured as calibration: irrespective of whether predictions are truly accurate, are they what they predict? A predictive test calibrates well when, say, a 16% risk means that 16% of all people with a 16% risk will develop the disease and a 30% risk means that 30% of the people with a 30% risk develop the disease, and so on for all levels of predicted risk.

Calibration is essential in DTC testing as predictions are based on formulas and data obtained from the scientific literature that might not apply to customers of every age, sex, and ethnicity. The question is not whether predictions of DTC companies are well-calibrated, but by how much calibration is off. If they want to investigate the calibration of polygenic risk predictions, companies need well-designed research studies for every disease that they predicted, in a population that is comparable to their customer base. It is safe to assume that companies do not have these data and that the accuracy of DTC polygenic risk predictions is and remains unknown.

Calibration is essential, but not enough. If a test predicts, say, 15% risk for diabetes to all its customers and this was a correct estimate, then the test calibrates well but is uninformative: the test cannot distinguish people who are more likely to develop a disease from those who are less likely. A test that *discriminates* well predicts higher risks for people who are at high risk and lower for those at lower risk. If, for example, the average risk of breast cancer among women is 12%, a test that does not discriminate well predicts risks between 10% and 14% for all women, while a discriminative test might predict risks that range from 1% to 60%. When these latter predictions are well-calibrated, the test can inform who is at higher and lower risk.

Assessing discrimination requires the same studies that are not available, but, unlike calibration, it can be evaluated using simulated data. That is how we investigated the predictive ability of six polygenic risk scores offered by three different companies, two of which no longer exist (Kalf et al., 2014). We showed that the discriminative ability varied markedly between diseases, but not between companies. Yet, this did not mean that consumers would get the same risk predictions from different companies. For individual customers, risk predictions for the same disease varied widely because companies used different algorithms and different data from scientific studies to perform the risk calculations (Kalf et al., 2014). While companies can validly estimate the discriminative ability of their tests, to the best of my knowledge, none of them is disclosing this information.

Therefore, companies do not know how well their risk estimates calibrate and discriminate, which would indicate the *clinical validity* of their tests, but they often do refer to the accuracy of their tests. This accuracy is the *analytical validity* of the test, which tells how accurately the tests genotype the DNA. In other words, when a test tells your genotype on SNP rs7903146 in the *TCF7L2* gene is CT, is it really CT? Analytic validity is always relevant, but particularly when a test informs about single DNA variants, such as the presence or absence of genetic mutations. Excellent analytical validity is no guarantee that the test also has good calibration and discriminative ability. These are different qualities of the test.

Experts agree that the predictive value of polygenic risk is limited, but expect that it will further improve when more SNPs are identified and included in the predictions

(Krier, Barfield, Green, & Kraft, 2016; Mihaescu et al., 2009). Yet, for most diseases such as heart disease and type 2 diabetes, it remains uncertain whether polygenic tests will achieve similar predictive ability as tests based on clinical risk factors that are freely available on the internet. Then why do consumers buy these tests?

INTENTIONS, UNDERSTANDING, AND PSYCHOLOGICAL IMPACT

Intentions

People are interested in broad personal genome services for various reasons. They may just be curious about their DNA, interested in learning about their genetic risks of diseases, or in finding opportunities to improve their health (Annas & Elias, 2014; Goldsmith, Jackson, O'Connor, & Skirton, 2012; Gollust et al., 2012). Since 23andMe halted the marketing of its health-related genetic testing and ancestry testing had become the sole and selling part of its personal genome services, the question is whether people will still be interested in buying its service if ancestry testing was not included. Consumers may have lost their interest now that it is wider known that the predictive ability of polygenic risk scores is limited. The Helix marketplace for genetic tests currently offers 35 genome-based information services, five of which are health tests (accessed: October 21, 2018). Only one of five informs about polygenic risks in addition to carrier status for genetic mutation and pharmacogenetic tests; the four others offer carrier testing only.

To understand the impact of genetic testing, it is important to know the purpose of testing: why does a customer buy the test (Borry & European Society of Human Genetics, 2010)? The response to testing may differ depending on whether a consumer orders a test because they have a specific question or whether they are just curious to peek into their DNA. The purpose of testing sets expectations about the test results, frames their interpretation, and affects the emotional and behavioral response to learning about risk. It does make a difference whether people with a family history of Alzheimer's disease buy a DTC genetic test to learn about their *APOE* status and people without a family history who receive their *APOE* test results complementary with their ancestry test.

Understanding

Scientific studies that investigated how well customers understand polygenic risk (Gordon et al., 2012; Kaphingst et al., 2012) showed that they do understand that "multifactorial disease" means that DNA is not deterministic and they adequately recalled risk results.

People seem to have an adequate understanding of the test and test results, but this may strongly depend on the questions asked. In one survey, more than 95% of the respondents correctly knew that people "can have a gene for a condition but not develop that condition" and that "common health conditions are caused by genes in combination with lifestyle and environmental factors" (Gollust et al., 2012). Yet, the first statement is inaccurate (all people have those genes, but only some have a mutation in the gene that causes the disease), and the second is uninformative as the relative contribution of the various factors differs substantially between common diseases. There is no consensus about what consumers need to know about DTC testing to make an informed decision about buying a test and what they need to understand to interpret the test results appropriately and to decide whether and how to act upon them.

Another study showed that lay people and genetic counselors had comparable interpretations of the risk assessments, but they had different thoughts about how helpful the results were. Most lay people, who had heard about DTC genetic testing but not bought a test, thought the tests were helpful, whereas most counselors thought they were unhelpful (Leighton, Valverde, & Bernhardt, 2012). The PGen study, a large study among customers of two personal genome companies, showed that people were less certain about their level of knowledge at six months *after* testing (Carere et al., 2016). Participants showed a high level of genetic knowledge but felt less certain about understanding it after testing. "I am confident in my ability to understand information about genetics" was (strongly) endorsed by 80% of the respondents at baseline and 62% at follow-up, and "I have a good idea about how genetics may influence risk for disease generally" from 74% to 64%.

These observations show that consumers may have adequate knowledge about genetics, but not about the impact and implications of the risk predictions. They may have a superficial knowledge about testing but doubt when asked about details. These observations raise and will continue to raise questions about what people need to know when buying a DTC test and what they can understand. Is a basic understanding of genetics enough? What do they need to know about the predictive ability of tests, the opportunities for disease prevention, and the implications of knowing they are at increased genetic risk? It cannot be assumed that people will understand, even if they say they do.

Psychological wellbeing

The several scientific studies on the impact of DTC genetic testing suggest that receiving risk information does not seem to increase anxiety and distress (Bloss, Schork, & Topol, 2011; Stewart, Wesselius, Schreurs, Schols, & Zeegers, 2018). In the worst-case scenario, people may experience anxiety shortly before and after testing, but these feelings restore over time. These observations sound comforting, but they should be no surprise as the genetic tests that were studied had limited predictive ability.

Whether the psychological impact is also minimal for DTC tests that can reveal substantially increased risks, such as *APOE* testing for Alzheimer's disease and *BRCA* testing for breast cancer, is another question. Studies suggest that people handle these results well too (Francke et al., 2013; Green et al., 2009), but studies like these often had a significant proportion of people who did not want to participate in the study, suggesting that the psychological burden may be underestimated (Janssens, 2015). The emotional response may further be underestimated as many of these participants knew that they could be at risk because of their family history. People who have a family history of Alzheimer's disease may respond differently to receiving APOE test results than those who are unprepared to the possibility of being at increased risk (Green et al., 2009).

The absence of adverse psychological impact cannot be assumed for future DTC tests. The current offer of polygenic risk has an unimpressive predictive ability that should not expect to elicit adverse emotional reactions (Janssens, 2015; Oliveri, Howard, Renzi, Hansson, & Pravettoni, 2016); test results may lead to more anxiety when tests become more predictive. It is also possible that over time people may be more familiar with DTC testing and its opportunities and limitations, which may make them better prepared to deal with adverse and unexpected results. Also, the quality of the assessment of effects on psychological wellbeing can be expected to improve as well, with better assessments of understanding, expectations, psychological wellbeing before and after, and intentions to act on health risk information.

In the critical appraisal of studies on attitudes, knowledge, and psychological impact, it is important to realize that most studies and surveys were conducted among early adopters of DTC testing and volunteers. These people tend to be more often Caucasian, healthier, have higher education, and higher income (Stewart et al., 2018). In a 2013 national survey, "only" 36% of the participants were aware of DTC genetic testing, which underscores the bias in these studies. Awareness was higher among individuals with a higher income, higher numeracy, and among information seekers (Agurs-Collins et al., 2015). Continued research is needed on the impact of DTC genetic testing, as what worked for the early adopters may not work for all.

UTILITY OF TESTING

Experts disagree about whether DTC genetic testing can be beneficial for disease prevention. Studies that investigated whether people change their health behavior after learning about their genetic risks have, at best, yielded mixed results. Most studies show no behavior change or only in the short term (Hollands et al., 2016). The lack of behavioral impact should not be surprising given that these polygenic risk tests were based on only a few SNPs and had low predictive ability. It is very possible that risk tests can motivate customers to adopt a healthier lifestyle when the tests can predict risks that are substantially increased. Whether genetic risk tests can become that predictive and whether substantial risks might lead to increased motivation for behaviors that are hard to change remains worth investigating.

The observation that studies did not observe behavioral change at large does not rule out that among the participants were people for whom the genetic risks did increase motivation (Stewart et al., 2018). One size does not fit all, and genetic testing might work for some but not for many others. The question then is: can we find out who might benefit from genetic risk information and, also, who might have adverse emotional reactions? Whether it is in the companies' interests to identify these people before selling the tests may depend on whether it helps them sell more or fewer tests.

Because of the lack of behavioral change and the limited predictive ability, DTC genetic testing is often criticized for having no utility, but proponents disagree. They argue that the DTC genetic testing industry is built on the premise that it is up to the consumer to determine whether a test has *personal* utility (Foster, Mulvihill, & Sharp, 2009; Grosse, McBride, Evans, & Khoury, 2009). A commercial DNA test does not need to have *clinical* utility as it is up to the customer to think what they find useful. Clinical utility means that the results of the test should help decide about medical treatment, but a test without clinical utility can still provide information that the consumer might value, even if the test has no informative value. While it may be up to the consumer to decide, this does raise ethical concerns. The consumer does not know whether a test provides valuable information when they are not informed about the discriminative ability of the test. In other words, when they do not know what test results other customers receive, it cannot be ruled out that they all receive the same risk information. In that case, the test cannot discriminate between people at higher and lower risk and is essentially a fake test.

WHAT ARE THE CONCERNS

The DTC genetic testing industry is booming, again, and likely here to stay, but the question is, what tests will be offered in the future? It is reasonable to assume that tests for carrier status will remain to be offered and ultimately expand to include all known recessive and dominant mutations. These tests are of interest to prospective parents who want to minimize the risk of passing mutations to their offspring, and to others who wish to know whether they will develop or are at substantially increased risk for a late-onset hereditary disease. Such broad mutation tests may also have undesirable uses, including antiselection by insurance companies and employers, for which a timely ethical discussion and legal protection remains needed.

Whether the polygenic risk tests are still offered in the future is uncertain. It seems reasonable to expect that they will continue to be offered as part of broader genomic services, but whether they will become marketable products on their own may depend on whether their predictive ability will further improve. The predictive ability of polygenic risk prediction is limited for most diseases, except when one or more SNPs has a strong impact on disease risk. Since these strong effects are the easiest to discover in scientific studies, it seems safe to assume that all exceptions are already known. Educating consumers and health professionals about the (lack of) predictive value of polygenic risk tests has a high priority.

Another concern about DTC genetic risk tests is the lack of transparency about how the risks are calculated. The DTC industry exists on the premise that the customer should make their own decision about these tests and whether they think testing is useful, but companies do not disclose technical details of the tests. The algorithms for their risk predictions, nutrition, and exercise recommendations are proprietary. It is not possible to make an informed decision about whether a test has personal utility. One can argue that consumers lack the knowledge to assess the predictive ability—which might be true—but without insight in the algorithms, experts are unable to evaluate on their behalf. The quality and predictive ability of the tests may converge at one point, but we are not there yet.

When tests become more complex, the interpretation of the results will become a challenge to the extent that it should be questioned whether lay people, not educated in genetics, can interpret the results without counseling by an expert. For polygenic risk, we may also reach a point that genetic education alone is not enough and that even geneticists and genetic counselors need to be trained in complex statistics to make sense of the risk scores. The correct interpretation of risk is already a challenge for the polygenic scores that use advanced statistical modeling to compute risk scores based on millions of SNPs (Khera et al., 2018).

The problems with interpretation are not only relevant for polygenic risk; the correct interpretation is also a challenge for the negative results of carrier testing. Geneticists and physicians have raised concerns about the *BRCA* test of 23andMe (Gill, Obley, & Prasad, 2018). This test checks only three of the thousands of mutations that are known in the two *BRCA* genes—three mutations that are predominantly found among women of Ashkenazi Jewish ancestry. When a woman carries the mutation, then she has a substantially increased risk of breast and ovarian cancer, but what it means when she does not carry the mutation depends on her background. When the woman is from an Ashkenazi Jewish family in which one of the three mutations causes breast cancer, then the absence of the mutation means that she will not develop the disease that runs in her family. For non-Ashkenazi Jewish women with a family history of breast cancer, the absence of a mutation may give false reassurance as the disease in her family was likely caused by another mutation than those three, to begin with. Companies often explain what it means when they find a mutation, but provide only sparse information when they do not find one.

When DTC genetic tests are here to stay, training of healthcare professionals is crucial (Borry et al., 2018). Their understanding of genetic tests is often limited, and they are not prepared to answer questions from their patients. We also need to put protocols and regulations in place to secure a responsible referral from companies to health care. When genetic tests are not predictive enough, then the results lead to many unnecessary referrals and follow-up confirmatory testing. Consumers should only be referred when the results of their genetic tests are such that the physician would have acted upon the same results if they had performed the test (Bijlsma et al., 2014). The costs of unnecessary follow-up testing caused by the companies should not be paid by health care.

REFERENCES

Agurs-Collins, T., Ferrer, R., Ottenbacher, A., Waters, E. A., O'Connell, M. E., & Hamilton, J. G. (2015). Public awareness of direct-to-consumer genetic tests: Findings from the 2013 U.S. Health Information National Trends Survey. *J Cancer Education, 30*(4), 799–807.

Annas, G. J., & Elias, S. (2014). 23andMe and the FDA. *New England Journal of Medicine, 370*(11), 985–988.

Bijlsma, M., Rendering, A., Chin-On, N., Debska, A., von Karsa, L., Knopnadel, J., ... Janssens, A. C. (2014). Quality criteria for health checks: Development of a European consensus agreement. *Preventive Medicine, 67*, 238–241.

Bloss, C. S., Schork, N. J., & Topol, E. J. (2011). Effect of direct-to-consumer genomewide profiling to assess disease risk. *New England Journal of Medicine, 364*(6), 524–534.

Borry, P., & European Society of Human Genetics. (2010). Statement of the ESHG on direct-to-consumer genetic testing for health-related purposes. *European Journal of Human Genetics, 18*(12), 1271–1273.

Borry, P., Bentzen, H. B., Budin-Ljosne, I., Cornel, M. C., Howard, H. C., Feeney, O., ... Felzmann, H. (2018). The challenges of the expanded availability of genomic information: An agenda-setting paper. *Journal of Community Genetics, 9*(2), 103–116.

Carere, D. A., Kraft, P., Kaphingst, K. A., Roberts, J. S., Green, R. C., & Grp, P. G. S. (2016). Consumers report lower confidence in their genetics knowledge following direct-to-consumer personal genomic testing. *Genetics in Medicine, 18*(1), 65–72.

Foster, M. W., Mulvihill, J. J., & Sharp, R. R. (2009). Evaluating the utility of personal genomic information. *Genetics in Medicine, 11*(8), 570–574.

Francke, U., Dijamco, C., Kiefer, A. K., Eriksson, N., Moiseff, B., Tung, J. Y., & Mountain, J. L. (2013). Dealing with the unexpected: Consumer responses to direct-access BRCA mutation testing. *PeerJ—the Journal of Life and Environmental Sciences, 1*, e8.

Gill, J., Obley, A. J., & Prasad, V. (2018). Direct-to-consumer genetic: Testing the implications of the US FDA's first marketing authorization for BRCA mutation testing. *JAMA—Journal of the American Medical Association, 319*(23), 2377–2378.

Goldsmith, L., Jackson, L., O'Connor, A., & Skirton, H. (2012). Direct-to-consumer genomic testing: Systematic review of the literature on user perspectives. *European Journal of Human Genetics, 20*(8), 811–816.

Gollust, S. E., Wilfond, B. S., & Hull, S. C. (2003). Direct-to-consumer sales of genetic services on the Internet. *Genetics in Medicine, 5*(4), 332–337.

Gollust, S. E., Gordon, E. S., Zayac, C., Griffin, G., Christman, M. F., Pyeritz, R. E., ... Bernhardt, B. A. (2012). Motivations and perceptions of early adopters of personalized genomics: Perspectives from research participants. *Public Health Genomics, 15*(1), 22–30.

Gordon, E. S., Griffin, G., Wawak, L., Pang, H., Gollust, S. E., & Bernhardt, B. A. (2012). "It's not like judgment day": Public understanding of and reactions to personalized genomic risk information. *Journal of Genetic Counseling, 21*(3), 423−432.

Green, R. C., Roberts, J. S., Cupples, L. A., Relkin, N. R., Whitehouse, P. J., Brown, T., ... Group, R. S. (2009). Disclosure of APOE genotype for risk of Alzheimer's disease. *New England Journal of Medicine, 361*(3), 245−254.

Grosse, S. D., McBride, C. M., Evans, J. P., & Khoury, M. J. (2009). Personal utility and genomic information: Look before you leap. *Genetics in Medicine, 11*(8), 575−576.

Hollands, G. J., French, D. P., Griffin, S. J., Prevost, A. T., Sutton, S., King, S., & Marteau, T. M. (2016). The impact of communicating genetic risks of disease on risk-reducing health behaviour: Systematic review with meta-analysis. *The BMJ, 352*, i1102.

Janssens, A. C. (2010). Why realistic test scenarios in translational genomics research remain hypothetical. *Public Health Genomics, 13*(3), 166−168, discussion 169−170.

Janssens, A. C. (2015). The hidden harm behind the return of results from personal genome services: A need for rigorous and responsible evaluation. *Genetics in Medicine, 17*(8), 621−622.

Janssens, A. C. J. W. (2014) How FDA and 23andMe dance around evidence that is not there. *Huffington Post*.

Kalf, R. R., Mihaescu, R., Kundu, S., de Knijff, P., Green, R. C., & Janssens, A. C. (2014). Variations in predicted risks in personal genome testing for common complex diseases. *Genetics in Medicine, 16* (1), 85−91.

Kaphingst, K. A., McBride, C. M., Wade, C., Alford, S. H., Reid, R., Larson, E., ... Brody, L. C. (2012). Patients' understanding of and responses to multiplex genetic susceptibility test results. *Genetics in Medicine, 14*(7), 681−687.

Khera, A. V., Chaffin, M., Aragam, K. G., Haas, M. E., Roselli, C., Choi, S. H., ... Kathiresan, S. (2018). Genome-wide polygenic scores for common diseases identify individuals with risk equivalent to monogenic mutations. *Nature Genetics, 50*(9), 1219−1224.

Krier, J., Barfield, R., Green, R. C., & Kraft, P. (2016). Reclassification of genetic-based risk predictions as GWAS data accumulate. *Genome Medicine, 8*, 20.

Lachance, C. R., Erby, L. A. H., Ford, B. M., Allen, V. C., & Kaphingst, K. A. (2010). Informational content, literacy demands, and usability of websites offering health-related genetic tests directly to consumers. *Genetics in Medicine, 12*(5), 304−312.

Leighton, J. W., Valverde, K., & Bernhardt, B. A. (2012). The general public's understanding and perception of direct-to-consumer genetic test results. *Public Health Genomics, 15*(1), 11−21.

Mihaescu, R., van Hoek, M., Sijbrands, E. J., Uitterlinden, A. G., Witteman, J. C., Hofman, A., ... Janssens, A. C. (2009). Evaluation of risk prediction updates from commercial genome-wide scans. *Genetics in Medicine, 11*(8), 588−594.

Oliveri, S., Howard, H. C., Renzi, C., Hansson, M. G., & Pravettoni, G. (2016). Anxiety delivered direct-to-consumer: Are we asking the right questions about the impacts of DTC genetic testing? *Journal of Medical Genetics, 53*(12), 798−799.

Ramos, E., & Weissman, S. M. (2018). The dawn of consumer-directed testing. *American Journal of Medical Genetics Part C—Seminars in Medical Genetics, 178*(1), 89−97.

Stewart, K. F. J., Wesselius, A., Schreurs, M. A. C., Schols, A., & Zeegers, M. P. (2018). Behavioural changes, sharing behaviour and psychological responses after receiving direct-to-consumer genetic test results: A systematic review and meta-analysis. *Journal of Community Genetics, 9*(1), 1−18.

CHAPTER 7

Assessing the Psychological Impact of Next-Generation Sequencing Information in the Clinic: An Attempt to Map Terra Incognita?

Marc S. Williams
Genomic Medicine Institute, Geisinger, Danville, PA, United States

INTRODUCTION

In the 2015 State of the Union address, President Obama called for investment in a large-scale precision medicine initiative—an initiative that subsequently garnered bipartisan support and initial funding to the tune of 200 million dollars[1] (Collins & Varmus, 2015). This initiative, renamed All of Us, has gone through an extensive planning process and in 2018 was beginning to enroll participants. As stated on the website, "The All of Us Research Program seeks to extend precision medicine to all diseases by building a national research cohort of one million or more U.S. participants." It is described as, "a participant-engaged, data-driven enterprise supporting research at the intersection of human biology, behavior, genetics, environment, data science, computation and much more to produce new knowledge with the goal of developing more effective ways to treat disease." The National Institutes of Health (NIH), which sponsors the project, notes that the timing for this is propitious given the convergence of several factors which include:

- Wide adoption of electronic health records (EHR);
- Costs associated with genomic analysis have dropped significantly;
- Data science has matured to the point that big data analysis is feasible;
- Health technologies are increasingly mobile;
- Americans are more engaged in managing their health conditions and interested in participating in research.

This initiative was committed to significant stakeholder engagement prior to launch, not the least of which was participation by representatives of the participants likely to enroll in the project. The choice of "All of Us" for the project reflects the

---

[1] Retrieved June 10, 2018, from https://www.whitehouse.gov/precision-medicine.

Clinical Genome Sequencing
DOI: https://doi.org/10.1016/B978-0-12-813335-4.00007-6

intention to have representation from anyone in the United States that wants to participate, and efforts to engage with groups traditionally underrepresented in research. While genomics will be an integral part of the project in the future, there is a recognition that this is not the only type of information that is of relevance to health and disease. Concerted efforts are underway to capture rich EHR data through patient-controlled download of medical record data, data from wearable sensors and devices, facilitated through novel interfaces such as Sync for Science.[2]

The All of Us project intends to return results to participants including findings from sequencing that will be done as part of the project. To determine the best approach for this, a Genomics Working Group was convened. This group released its initial report on December 5, 2017.[3] The group evaluated three technologies that could be used for the participants, whole genome genotyping (WGG), whole exome sequencing (WES), and whole genome sequencing (WGS), and assessed their potential application for the program. The group recommended that WGG and WGS be considered for the project on the basis of the value of the information available to be returned to participants and for research where WGS was determined to have sufficient incremental value over WES to justify the additional expenditure of resources needed to generate the sequence. The group also recommended a pilot phase involving ~5% of the anticipated total cohort deeming that to have "...sample size sufficient for testing pipelines, evaluating data types, capturing variants across the diverse participant base of the program, and assessing return-of-information strategies."

Given the large-scale investment by the NIH in All of Us and other sequencing projects, it is timely to consider the psychological implications of next-generation sequencing (NGS) for patients and family members. This chapter will address this important topic across a broad range of indications for sequencing outside of the research setting. The use of sequencing in the clinic is still emerging and the study of psychological implications in response to sequencing results is very early, meaning that empiric evidence to support the assertions presented are lacking. To mitigate this limitation, in addition to citing emerging evidence that directly addresses use of NGS in the clinic, extrapolation from research using more traditional genetic testing technologies will be used where applicable. For areas with little or no evidence, informed speculation by the author will be included along with relevant arguments to support the conclusion and define limits to the speculation. Therefore, this chapter should not be viewed as a definitive guide to the psychological impact of NGS, but rather, as a first attempt at mapping terra incognita. The hope is that this may help researchers to

---

[2] Retrieved June 10, 2018, from https://www.healthdatamanagement.com/news/sync-for-science-helping-patients-share-ehrs-with-researchers.

[3] Retrieved June 10, 2018, from https://allofus.nih.gov/sites/default/files/gwg_final_report.pdf.

generate research questions of relevance to the topic to generate the empiric evidence needed to support the anticipated implementation of genomic medicine.

Before diving into the topic at hand, it is necessary to provide some background to the reader to orient them to two important issues: the profusion of terms that are inconsistently used in the literature and lay press, and a description of the different applications of NGS in the clinic, as the potential for psychological distress will likely vary depending on the indication. The following section provides the author's perspective on these two issues.

BACKGROUND

Definition of terms

There is a profusion of terms being used to describe activities that propose to combine NGS data with other data to improve the care of patients. These include the historical term genetics and emerging terms that include genomics, personalized (or individualized) medicine, precision medicine and, most recently, precision health. While some treat these terms as interchangeable, there are important distinctions that should be recognized and used to inform when the different terms should be used. There is no agreed-upon lexicon, so what follows in the author's assessment and evaluation of the terms. These definitions will be used in the text as different applications of NGS are discussed.

The first to examine is the difference between genetic and genomic. The World Health Organization defines genetics as "the study of heredity," while genomics is defined as "the study of genes and their functions, and related techniques" (Genomics and World Health, 2004). The distinction that is most commonly made is that genetics concerns itself with the functioning and composition of a single gene, whereas genomics addresses all genes and their interrelationships. This general distinction in biology is a bit clearer when applied to medicine. The specialty of clinical genetics has emerged, been identified, and primarily limited to the diagnosis of so-called single-gene disorders, also known as Mendelian disorders as for the most part they follow the rules of inheritance reported by Gregor Mendel in the 19th century. Genomic medicine, in contrast, is defined by the National Human Genome Research Institute in 2015 as, "an emerging medical discipline that involves using genomic information about an individual as part of their clinical care (e.g., for diagnostic or therapeutic decision-making) and the health outcomes and policy implications of that clinical use."[4] Finally, while neither definition includes a specific reference to family history, it is generally assumed that family history of disease is included in both genetic and genomic medicine.

---

[4] Retrieved June 10, 2018, from http://www.genome.gov/27552451.

A term that was in common usage until relatively recently was personalized medicine. Is personalized medicine really a new or, more importantly, a finely specified, concept? Most providers would say that they have always practiced personalized medicine. The history of medicine is replete with exhortations to personalize care. Hippocrates emphasized the observation of the individual patient, documentation of signs and symptoms, and tailoring treatment to the individual based not only on the signs and symptoms, but also the values of the patient. Indeed, this probably represents the first recorded reference to patient-centered care.

Some consider personalized medicine to be synonymous with genomic medicine, but here the term genomic medicine may be more meaningful. Given the profusion of genetic and genomic discoveries in the past decade it is hard to argue against the potential for this new information to impact medicine. There is a sense from some individuals that this knowledge is so fundamental to our understanding of health and disease that it will relegate other information to a second-class status. The image of our genetic code on a credit card that will be inserted into a computer in the doctor's office so that our personal data can be used to direct care is compelling, although the complexity of the science suggests this vision is unlikely to be completely realized in the near term. Still, it is important to recognize that the use of biomarkers—genes or other biological substances—to personalize therapy is not new.

The definition of personalized medicine proposed by Pauker and Kassirer (1987) while preceding the Human Genome Project has much to recommend its use. "Personalized medicine is the practice of clinical decision-making such that the decisions made maximize the outcomes that the patient most cares about and minimizes those that the patient fears the most, on the basis of as much knowledge about the individual's state as is available." There are three key points captured by this definition. First, is the focus on the outcomes of care. Second, is the central role of the patient in defining what outcomes, positive or negative, are of most importance; indeed, the authors anticipated the emphasis on patient-centered care by two decades. Third, notice that the words "genetic" or "genomic" do not appear in the definition. The authors emphasize the importance of using "...as much knowledge about the individual's state as is available." There are no assumptions that genetic or genomic information is superior to other information in caring for the patient. Only if evidence demonstrates that genetic information is superior should it be treated as such. Consider a commonly performed medical procedure, blood transfusion. Prior to 1930 this was a very dangerous procedure that could result in the death of the patient from a transfusion reaction that could not be predicted. In 1930 the discovery of ABO blood groups by Karl Landsteiner allowed safe transfusion of blood products for the first time. ABO groups are genetically determined, but no one would suggest we forgo blood typing and rely on genetic markers to determine which blood to transfuse. While the focus of this chapter will be on genomics, it is important not to lose sight of this important admonition, as the psychological implications are more likely to be influenced by the content of the information, rather than the type of data used.

Although the concept of personalized medicine is not new, evidence is accumulating that the practice of medicine is in the process of being transformed. The realization of personalized medicine to this point has been empiric and dependent on how much knowledge and experience the individual provider has. This has led to care that has high variability and less than optimal outcomes. Christensen, Grossman, and Hwang (2009) refer to this as "intuitive medicine" and define it as "...care for conditions that can be diagnosed only by their symptoms and only treated with therapies whose efficacy is uncertain." The focus in care is moving from the intuitive to the precise, but that, of course, is feasible only to the extent that the knowledge needed to do it exists. The same authors define precision medicine as, "...the provision of care for diseases that can be precisely diagnosed, whose causes are understood, and which consequently can be treated with rules-based therapies that are predictably effective." It is at the intersection of personalized and precision medicine where the opportunity exists to optimize patient care. Precision health is the newest term to be promoted. At Geisinger, the author's institution, we have adopted this term for our large-scale sequencing project, the MyCode Community Health Initiative (Carey et al., 2016), as we view this as a population health effort. The term, precision health, emphasizes prevention while encompassing the interventions inherent in precision medicine, thus broadening the application of NGS to include population-based initiatives (Williams et al., 2018).

INDICATIONS FOR NEXT-GENERATION SEQUENCING

NGS is being proposed as an efficient approach for a diverse set of clinical scenarios because of its ability to query multiple genes and variants in a single assay. The cost of testing single genes sequentially compared to a next-generation panel, or WES favors the latter once more than two or three genes of interest are relevant to the condition under study. Of course, this generates information on many more genes than might be of interest to the clinician or patient—a trade-off which will be discussed in detail below. For this reason, NGS is increasingly proposed as a first-line test. The different reasons for testing (called test indications in medicine) carry with them different levels of risk and concern as perceived from the perspective of patients, families, and providers. These reasons are defined in detail in Chapter 2, Genome Sequencing and Individual Responses to Results, but since these indications will be used to organize the subsequent discussion, a brief explanation of each is appropriate to orient the reader.

Diagnostic testing

Diagnostic testing is performed to answer the question of what is causing a specific set of signs, symptoms, laboratory, or imaging findings present in a patient. NGS-based approaches such as gene panels and WES have been used for a variety of clinical indications where the likelihood of genetic disease is high (Genetic Testing Registry Search). Evidence is emerging that supports the use of NGS-based approaches as a first-line test for several indications based on diagnostic yield and decreasing sequencing costs, although given methodologic limitations, more rigorous evaluations are needed (Schwarze, Buchanan, Taylor, & Wordsworth, 2018; Stark et al., 2018). This subject is covered in detail in Chapter 2, Genome Sequencing and Individual Responses to Results.

Pharmacogenomic testing

The pharmacogenome is the subset of the genome that codes for proteins involved in drug transport, metabolism, and response (including adverse events).[5] While some pharmacogenomic (PGx) information can be extracted from WGS and WES data, the most effective PGx testing approach uses NGS panels constructed for this purpose.

Somatic testing

One of the most exciting developments in the treatment of cancer has been the identification of molecular signatures that can be targeted by specific therapeutic agents. This indication, which is now called precision oncology (Bode & Dong, 2017) has been used for several years and is informed by genomic information. While still early in development, molecular tumor profiling has demonstrated success in non-small cell lung cancer (NSCLC) (Schallenberg, Merkelbach-Bruse, & Buettner, 2017) and other cancers to a lesser degree (Chae et al., 2017). While the evidence for the efficacy of molecularly targeted therapies remains immature, there are early indications that the application in advanced cancer patients improves survival and lowers weekly healthcare costs (Haslem et al., 2018). NGS tumor profiling usually provides information restricted to the tumor, but some approaches such as paired WES for normal and tumor tissue can yield information on germline genetic information that may be of relevance to the patient, including additional findings discussed below.

Population screening

NGS can be applied to unselected populations of patients as a screening test to identify risk for genetic disease in the absence of an indication (i.e., signs or symptoms of a disease). At present, this is being conducted in the research setting (e.g., MyCode,

---

[5] Retrieved June 10, 2018, from https://www.pharmgkb.org/whatIsPharmacogenomics.

ClinSeq, UK) (Carey et al., 2016)[6] and is not considered to represent a clinical standard of care. This indication carries with it the highest potential uncertainty about the meaning of the result—an issue that is discussed in detail below.

Incidental, secondary, or additional findings

This area of NGS results is of such importance that it is the focus of Chapter 2, Genome Sequencing and Individual Responses to Results, and Chapter 12, Opportunistic Genomic Screening: Ethical Exploration, and will not be discussed in this chapter.

Each of these indications has a unique set of issues that impact the clinical and psychological outcomes of patients undergoing testing. The remainder of the chapter will explore these in detail.

PSYCHOLOGICAL IMPACT OF DIFFERENT TYPES OF NEXT-GENERATION SEQUENCING

In general, the psychological impact of genetic testing, even for serious, and currently untreatable conditions such as Alzheimer's disease, has been modest (Graves et al., 2012; Picot et al., 2009; Rahman et al., 2012). While this provides some reassurance, these studies focused on targeted testing either for an indication (breast, ovarian cancer) or to evaluate risk for a specific condition. The studies did not address either perceived differences in the application of NGS technologies which, by their nature, may return a broader range of results, are used to screen across multiple conditions, or return results not related to the indication for the testing. In this section, the psychological impact of the different types of NGS testing will be assessed. The chapter will be organized around the taxonomy of medical uncertainties proposed by Han, Klein, and Arora (2011) and discussed in detail in Chapter 5, Uncertainties in Genome Sequencing.

DIAGNOSTIC TESTING

The most common use of sequencing is for diagnostic purposes. NGS (exome, genome, and panel) has revolutionized the evaluation of children with complex conditions suspected to be genetic. Prior to the introduction of NGS the diagnostic yield from testing was around 15%. Sequencing has at least doubled the diagnostic rate in these complex children, with some conditions (the epileptic encephalopathies and

---

[6] Retrieved June 10, 2018, from https://www.genome.gov/20519355/clinseq-a-largescale-medical-sequencing-clinical-research-pilot-study/; Retrieved from June 10, 2018, https://www.genomicsengland.co.uk/the-100000-genomes-project/.

deafness) having much higher diagnostic rates (Córdoba et al., 2018; Lin et al., 2012). Making a diagnosis of a specific condition allows changes in medical care that support condition-specific management. Studies that have evaluated changes in medical management following a diagnosis, confirmed that the result alters medical management in a significant proportion of patients (Sands & Choi, 2017; Stark et al., 2018; Vissers et al., 2017), particularly when rapid exome sequencing is applied to critically ill newborns (Meng et al., 2017; Smith, Willig, & Kingsmore, 2015).

The psychological impact of single-gene genetic testing has been well established, at least in the context of cancer susceptibility. The evidence indicates that for patients receiving a negative result, pretest distress is stable or lessens, while those receiving a positive result typically have an increase in distress that returns to baseline usually over a 6—12-month period of time (Athens et al., 2017).

The psychological issues associated with this testing indication have primarily focused on the impact on the patient and family related to diagnostic uncertainty, which fits in the category of scientific uncertainty. Many of these patients have experienced diagnostic odysseys that have lasted years, with the attendant impact on the patient and family. This was studied in the context of the Undiagnosed Diseases Network (McConkie-Rosell et al., 2018). The authors note that prior work on the psychological impact of the diagnostic odyssey cuts across all three domains of uncertainty. Establishing a diagnosis reduces the scientific uncertainty and is usually associated with a positive impact on psychological wellbeing, even in the case of a diagnosis associated with a poor prognosis, such as Alzheimer disease (Christensen, Roberts, Uhlmann, & Green, 2011). In this situation, the idea that knowing the diagnosis allows the patient and family to focus on appropriate care and life planning issues (practical and personal uncertainties) rather than trying to answer the diagnostic question. This is especially true in the newborn ICU, where rapid diagnosis of an untreatable genetic disorder allows transition from invasive care to palliation and comfort care (Meng et al., 2017; Smith et al., 2015). However, this may not be the case for all disorders as seen in an exploratory study on genetic testing in deafness that did not identify significant psychological differences in patients and families who received a diagnosis compared with those that did not (Oonk et al., 2018). There is also a body of literature looking at the psychological impact on parents of caring for a child with a diagnosed complex disease, that confirms significant and ongoing psychological issues across all three domains, though mostly within practical and personal. Peay et al. (2016) note that while significant distress is present in both diagnosed and undiagnosed patients, those parents with a child that is undiagnosed are more impacted, although a study of uncertain results in diagnostic trio sequencing for undiagnosed genetic disease suggests that the negative impact is modest and most parents did not express regret about undergoing testing for their children (Skinner et al., 2018).

Negative psychological impact can also occur. Lumish et al. (2017) presented data from a study of 232 patients who underwent testing for cancer predisposition risk using a NGS panel. Their study found that unaffected individuals with a family history of breast or ovarian cancer who received positive results were most significantly impacted by intrusive thoughts, avoidance, and distress. They also found that measures of distress were modestly elevated among patients who received a variant of uncertain significance (VUS), suggesting that ambiguous results also confer distress, a finding consistent with increased irreducible uncertainty in the scientific domain. This assessment is reinforced as there was significant confusion associated with the interpretation of the result. This is an important finding given that VUS are frequent among patients undergoing genetic testing with large panels of genes, and with the number of VUS increasing with increasing panel size (Lumish et al., 2017). In addition, the use of larger panels leads to the inclusion of genes for which the gene–disease associations are more tenuous. For example, the ClinGen expert gene curation panel looking at Brugada syndrome[7] identified 21 genes currently available on testing panels. However, evidence for a definitive gene–disease association was established for only one gene, *SCN5A* (Hosseini et al., 2018). This acts to increase scientific uncertainty for both the patient and provider, and can confound clinical decision-making. A smaller study done in the context of testing for inherited cardiomyopathy (Wynn, Holland, Duong, Ahimaz, & Chung, 2018) found similar results. Patients receiving a positive result had higher scores for intrusive thoughts, avoidance, and distress when compared to those with negative genetic test results. However, they also exhibited more formal planning about making health behavior and lifestyle changes, an indication that the patients were acting to reduce uncertainty in the practical and personal domains. Interestingly, the satisfaction with the decision to undergo testing was similar between those receiving positive and negative results, a finding also seen in the cancer panel testing study (Lumish et al., 2017). This suggests that the magnitude of the psychological distress is modest.

A negative test (that occurs more than half the time for most indications) may also have negative psychological implications (Skinner, Raspberry, & King, 2016; Werner-Lin et al., 2018). A negative finding does not reduce the scientific uncertainty associated with the underlying condition. Most patients and families experience emotional ups and downs associated with the hope that each new test will provide the answer, followed by the letdown when this test is negative. These up and down cycles take an emotional toll on patients and families. It is important to manage expectations at the time of testing to address reducible uncertainty in the practical and personal domains. One potential advantage of sequencing (more so exome/genome compared to panels)

[7] Retrieved June 10, 2018, from https://www.clinicalgenome.org/working-groups/clinical-domain/cardiovascular-clinical-domain-working-group/brugada-syndrome-gene-ep/#wg_ep_status.

is the potential to reanalyze the sequence periodically to take advantage of new knowledge about gene–disease associations. While this has not been routinely incorporated into clinical care (Ewans et al., 2018), if this is offered it is important to keep contact with the patient and family, both to identify new medical issues that could provide new approaches to interpretation, but also to let them know when reinterpretation has been done and what was found.

Based on the above, it is likely safe to extrapolate from more traditional genetic testing given the similarity of the indication. Additional large prospective studies, such as the Psychosocial Genomic Oncology Project (PiGeOn) (Best et al., 2018), are needed to more fully explore the psychosocial impact of diagnostic sequencing.

PHARMACOGENOMIC TESTING

PGx testing is emerging into clinical use. PGx testing can be done at the time a medication is ordered, or pre-emptively so that PGx information is available when a medication impacted by PGx is ordered. The latter approach is supported by evidence-based guidelines developed by the Clinical Pharmacogenomics Implementation Consortium (CPIC)[8] that provide information to the ordering provider about how the result should be interpreted and used to inform the choice and dose of medication. PGx can be divided into two categories of results: prediction of a severe adverse event (e.g., HLA-B*15:02 and Carbamazepine[9]) and results that provide information about the effectiveness of a given medication for the individual accompanied by recommendations for dose and/or alternative medications (e.g., *CYP2C19* poor metabolizers and Clopidogrel[10]). For the discussion in this section, only PGx results associated with drug response and not associated with other human disease will be considered.

Current pharmacy practice incorporates multiple patient-specific factors including age, weight, sex, liver and kidney function, allergies, and medication interactions into decisions to select and dose a new medication. Pharmacogenomics does not supersede this information, rather it adds to the clinical information to provide more precise direction to the clinician. An example of this is warfarindosing.org,[11] a publicly available warfarin-dosing algorithm that allows the clinician to input clinical information and PGx variant information for the genes that impact warfarin metabolism. This raises the question of whether PGx information is different from the clinical information beyond the perception that it may be different. While personalizing medications using clinical factors is the standard of care, it is not clear how aware patients are of this

---

[8] Retrieved June 10, 2018, from https://cpicpgx.org/.

[9] Retrieved June 10, 2018, from https://cpicpgx.org/guidelines/guideline-for-abacavir-and-hla-b/.

[10] Retrieved June 10, 2018, from https://cpicpgx.org/guidelines/guideline-for-clopidogrel-and-cyp2c19/.

[11] Retrieved June 10, 2018, from http://warfarindosing.org/Source/Home.aspx.

practice, beyond knowing that they will not be prescribed a medication they are allergic to. Most research on the perceptions of the use of PGx information has been focused on clinicians, which reflects this current paradigm. Therefore, it is difficult to extrapolate the implications of including PGx information with clinical factors from the patient perspective.

Patient-focused research on pharmacogenomics is limited and has been mostly centered on comprehension and education. At present, no study has focused on the psychological impact of returning PGx results to patients. Haga, Mills, and Bosworth (2014) discussed the potential psychological impact of returning PGx results, however, the focus of this perspective article was on the potential psychological impact of imperfect comprehension of the PGx report itself, which relates to uncertainty in the scientific domain. The authors speculate that risk perception resulting from the return of PGx results could impact medication adherence as proposed in the theory of planned behavior. However, they go on to note that at that time, there was almost no empirical study of these issues, and the one cited study had a very narrow scope (breastfeeding mothers taking codeine) and some methodologic issues. They do endorse the proposition in this chapter that PGx testing is different from other types of genetic testing and should be studied separately. A large study from the Mayo Clinic surveyed over 1000 patients enrolled in their pre-emptive PGx testing program, RIGHT (Olson et al., 2017). These individuals received the results of testing for the gene *CYP2D6*, that encodes for an enzyme important in the metabolism of multiple medications. The survey was primarily focused on the comprehension of the result, but some questions were included about the patient attitude toward receiving the result. These questions addressed four content areas: (1) whether they reported sharing or were planning to share their results with others, (2) whether they would encourage others to get PGx testing, (3) how useful they believed pharmacogenomic results will be, and (4) whether the use of pharmacogenomics would affect their medication compliance. While none of these content areas directly addresses issues of psychological impact, some inferences can be made. A total of 87% of respondents strongly or somewhat agreed to the statements, "Using my pharmacogenomic results when prescribing medications for me will improve my chances of getting a dose that is right for me," and "Using my pharmacogenomic results when prescribing medications for me will improve my chances of getting a medication that is right for me." Agreement with both of these statements is associated with a reduction in scientific uncertainty. Assuming that receiving the correct medication and dose are desirable (which seems highly likely), one can infer that the perceived benefit overwhelmed any psychological concerns about the genomic nature of the information. Nearly two-thirds of respondents indicated that they would encourage others to get PGx testing. This would imply that not only did they endorse the usefulness of the testing, but they did not experience excessive distress. In addition to not addressing psychological issues directly, there were other study limitations, including the use of a consented research population (more motivated to receive results) whose demographics were primarily Caucasian, female, and more highly educated. In addition, as with many studies in genomics, the results were based on what the participants planned to do with the results, rather than measuring actual behavior, although in contrast to other studies, the participants received results as opposed to responding to hypothetical results.

Geisinger conducted research developing patient-facing genomic test reports to support our large-scale precision health sequencing study, the MyCode Community Health Initiative (Carey et al., 2016). As part of this effort, and cognizant of the issues raised by Haga et al. (2014) we partnered with Geisinger patients and providers to develop paired patient- and provider-facing PGx test reports (Jones et al., 2018). Using user-centered design principles we conducted semistructured interviews to explore the content and attitudes about PGx information. Although not the primary focus of the project, the interviews included questions to explore patient and provider reactions to the information. While limited for the most part to a hypothetical reaction to different scenarios, the patients were uniformly positive about the importance of the information and did not express any potential concerns about receiving it. Two patients had experienced adverse events related to use of a statin and when presented with a PGx report that could have identified susceptibility prior to receiving the statin expressed how important this information would have been at the time of prescribing. As one patient said, *"That's right, if I would have had this, we could have saved a lot of grief and money, and you know, gone right to the top."* Interestingly, several providers were concerned that the report might cause patients to become upset, panicked, or confused. One provider stated, *"So, I can see patients panicking, and the first thing they do is call their physician who (a) if they weren't informed, is going to get very upset, and (b) is also going to get upset if they get the phone call and have to deal with it."* This apparent disconnect between patient and provider perceptions of the information is important to acknowledge and study, a point emphasized by Han et al. (2011) in discussing the importance of defining the locus of uncertainty (that is patient vs physician). This potential disconnect could raise uncertainty in patients based on provider information, issues involved in scientific and practical domains. These uncertainties are reducible and can be addressed through education, provider support, and access to providers with appropriate expertise. We agree with Haga that patient-focused empiric research on the reaction to PGx testing results is needed. Ideally this will involve participants that are broadly representative to account for the differences that may be present in different populations.

SOMATIC TESTING

One of the fastest-growing areas for NGS is somatic testing of tumors (Nakagawa and Fujita, 2018). Sequencing of tumors has provided insights into the molecular mechanisms that drive tumorigenesis and is beginning to transform the approach to treating some cancers, most notably NSCLC, a cancer that has historically been challenging to treat. Multiple targeted agents have been approved for use in NSCLC based on the molecular signature of the tumor and best practice guidelines support the sequencing of these tumors (Sholl, 2017). It is anticipated that the use of sequencing will be extended to more types of cancer and may become a standard of care as more information accumulates about the impact of molecularly targeted cancer therapy. A significant concern that is likely to produce psychological consequences for patients, at least in the United States, is the cost and access to molecular targeted interventions, uncertainties that fall in the practical and personal domains. While important this uncertainty will not be the focus of this section.

Sequencing of tumors primarily identifies somatic variation that, while providing important information about treatment and prognosis for the cancer, does not have broader implications for other disease risk or any concerns for family members. Tumor sequencing can identify germline variants that confer disease risk for the individual and their family members, however, these results are more closely aligned with diagnostic testing or additional findings, and hence will not be discussed in this section.

As with PGx testing, cancer therapies have always been determined by clinical factors combined with tumor characteristics, such as tissue of origin, histology, and other biomarkers, so sequencing information represents additional information that informs the therapeutic interventions. For the most part, patients are unaware of the detailed information derived from the tumor used to determine therapy and prognosis (Tzeng, Mayer, & Richman, 2010), and the addition of sequencing is unlikely to change this beyond high-level discussion of molecularly targeted therapies. While there are some studies looking at the psychological impacts of tests for gene expression used for risk stratification, there is a dearth of information about the psychological implications of presenting somatic sequencing information to patients. The only study that addresses this was an evidence assessment from 2007 that evaluated diagnostic strategies for hereditary nonpolyposis colorectal cancer, aka Lynch syndrome (Bonis et al., 2007). In this critical review, the authors noted that there were limited data about the potential harms of strategies to identify patients with Lynch syndrome. However, the information that was available for review suggested that testing was not associated with adverse psychological impact following formal counseling. They further reported that pretest genetic counseling had good efficacy in improving knowledge about Lynch syndrome and resulted in a high likelihood of proceeding with genetic testing. This supports the notion that this type of testing reduces scientific uncertainty regarding the intervention, while counseling and education can reduce uncertainty in the scientific, practical, and personal domains. This review is of some relevance to the topic of somatic sequencing in that tumor-based testing is recommended for the identification of individuals at risk for carrying a germline variant causing Lynch syndrome, however the assessment of psychological harms is limited to the discussion of diagnostic testing.

It seems reasonable to assume that the likelihood of psychological distress resulting from tumor sequencing is minimal, particularly in the context of coping with issues around the diagnosis and treatment of a life-threatening disease. However, a preliminary study by Gray et al. (2012) identifies some potential issues that suggest this assumption may not be correct. In this study, a cohort of patients with cancer was presented with information about the use of somatic testing for cancer and attitudes were elicited using semistructured interviews. Participants endorsed a high level of understanding of the nature and purpose of this testing. All participants reported positive aspects of somatic testing. A total of 71% identified some potential negative aspects to somatic testing, with about one-third expressing concerns about possible psychological harm. A subset of the cohort had undergone somatic testing during treatment for their cancer and they expressed much lower levels of concern compared with those who had not had somatic testing (21% vs 57%). While the potential for harm was acknowledged, it appears to be of modest concern, at least in comparison with the benefits, as 96% of participants stated they would take a somatic genetic test if offered.

Given the anticipated increase in use of tumor sequence information to guide care, including emerging technologies such as cell-free DNA and liquid biopsy, research regarding the psychological impact of this information on patients is desirable. The differences in perception between hypothetical and actual experience necessitate these studies be conducted with populations of actual patients. The second focus of the PiGeOn project referenced above is to assess the psychosocial impact on individuals whose tumors have undergone NGS (Best et al., 2018).

POPULATION SCREENING

At present, there is insufficient evidence for population screening using next-generation technologies. Currently, only a handful of single-gene disorders are screened for at the population level, and these are limited for the most part to populations at high risk for specific genetic disorders.

Population screening using NGS is being explored in several research settings. In the United States, the All of Us program is looking to enroll 1 million participants who will undergo WGG and WGS with return of results. This program has been informed by prior publicly funded research efforts, such as the CSER Consortium,[12] and the NIH ClinSeq study (Biesecker et al., 2009). Private population sequencing projects such as the Geisinger MyCode Community Health Initiative (MyCode)

[12] Retrieved June 10, 2018, from https://cser-consortium.org/.

(Carey et al., 2016; Williams et al., 2018) are also beginning to contribute information about population-level genome screening.

The psychological implications of learning nondiagnostic, unanticipated genomic sequencing results are being studied. However, it is important to appreciate a limitation of these studies. This research often requires opt-in consent, which implies that potential participants who have concerns about learning results for any reason will not participate in the project. This could lead to an underestimation of the psychological impact of learning results in the screening setting since only participants who consent will be included in the assessment. Presumably, if the individual had significant concerns about receiving such a result based on anticipated psychological response, they would not consent to participate. Two studies have gathered data from individuals who declined participation in large-scale population sequencing projects.

The CSER consortium addressed this in their survey of individuals who declined participation referenced above. In their survey of over 1000 individuals who declined participation, the most commonly cited reason was logistical complications, that is, personal situations that would not permit them to meet the requirements of the research study. Concerns about privacy and discrimination were expressed by 13%. While not typically considered as psychological, under the Han model, these could be included in the category of personal uncertainty. Only 8% cited concerns about psychological implications of the results as a reason for declining participation, but these were not explored further.

The most systematic and thorough analysis of psychological factors associated with NGS in a healthy population has come out of the ClinSeq project (Biesecker et al., 2009). The ClinSeq project is discussed extensively in Chapter 2, Genome Sequencing and Individual Responses to Results, but the author would be remiss not to highlight information from this seminal project that are of relevance to the subject of this chapter.

In 2015, ClinSeq published a series of four papers that explored different aspects of the psychological impact of sequencing including perceived ambiguity (Taber, Klein, Ferrer, Han et al., 2015), dispositional optimism (Taber, Klein, Ferrer, Lewis, Biesecker et al., 2015), current affect (Ferrer et al., 2015), and information avoidance (Taber, Klein, Ferrer, Lewis, Harris et al., 2015). These represent the first systematic studies of the psychological impact in a population sequencing setting. The findings of these papers are summarized briefly.

- Perceived ambiguity is a major component of most uncertainty constructs, including that of Han. In the ClinSeq study (Taber, Klein, Ferrer, Han et al., 2015), perceived ambiguity of sequencing results was defined as participant perceptions of the accuracy and interpretability of any potential future results. The authors' primary hypothesis was that the higher the perceived ambiguity of sequencing results, the more likely participants would negatively perceive the impact of the results, a phenomenon of ambiguity aversion. In general, ClinSeq participants expressed low levels of ambiguity about sequencing results while endorsing the high perceived value and benefit of the results, with low perception of potential harms. Additionally, it was noted that participants who perceived greater ambiguity about their sequencing results reported lower perceived efficacy, lower value, health benefits, and higher potential for harms, supporting the primary hypothesis.

- The second study (Taber, Klein, Ferrer, Lewis, Biesecker et al., 2015) examined dispositional optimism. The investigators tested the interactive effects of dispositional optimism and perceived risk among adults who reported intentions to learn their actual genome sequencing results and to change their lifestyle and health behaviors in response. Neither dispositional optimism nor comparative risk (that is the individual perception compared to other similar individuals) significantly impacted the intention to receive sequencing results based on absolute or experiential risk. However, a relationship between underlying dispositional optimism and comparative risk was identified as having an impact on receiving the results of carrier screening. Individuals with higher dispositional optimism expressed increased intention to make health behavior changes based on the sequencing results, irrespective of whether the result had associated medical actionability or not.

- The third study (Taber, Klein, Ferrer, Lewis, Harris et al., 2015) looked at the impact of current affect and anticipated negative affect on the intention to receive sequencing results. The study also examined the interaction between these participant characteristics and self-affirmation. It was thought important to include self-affirmation as there is evidence supporting the idea that self-affirmation in response to a perceived threat could impact anticipated negative affect. The results of the study showed that current affect had little impact on the intention to receive results. However, increased anticipated negative affect was associated with a decreased intent to receive results for both actionable and nonactionable conditions. In this study it did not appear that self-affirmation played a significant role in ameliorating the impact of anticipated negative affect.

- Information avoidance is an intriguing human behavior that represents a type of defense strategy to avoid information that is considered to be threatening, a behavior of some relevance for genetic test results (Taber, Klein, Ferrer, Lewis, Harris et al., 2015). In this study, the investigators examined whether individual differences in information avoidance predicted intentions to receive genetic sequencing results. As might be expected, individuals with higher tendencies toward information avoidance expressed less interest in receiving sequencing results. This was particularly acute in the scenario where the results did not have actions that could impact the development or progression of disease, a finding that emphasizes the interaction between information avoidance and perceived personal control. Additionally, individuals with low dispositional optimism or low self-affirmation when coupled with higher information avoidance showed lower intention to learn results.

While this information provides critical insights into the psychological impact of population sequencing, there are significant limitations to the ClinSeq project. The authors note that ClinSeq participants are likely favorably disposed to receiving sequencing results based on their choice to participate in the program. Participants have higher socioeconomic status and educational attainment than the United States population average. Participants were predominantly Caucasian and non-Hispanic. All of these factors could impact the generalizability of the findings. Finally, this series of papers tested hypotheses related to the intent to receive results as opposed to actual results received.

The Geisinger MyCode project has used extensive and ongoing community engagement to explore many issues related to the return of genomic results in a population setting. Prior to initiation of sequencing, several hundred Geisinger patients were consulted through interviews and focus groups to understand issues related to returning results. An overwhelming majority encouraged the project to return results to the participants (Faucett and Davis, 2016). Many expressed this as an expectation that results would be returned if there were implications for the participant's health, with a few expressing surprise that return of results is not routine for research studies. The few who raised concerns cited issues about genetic discrimination (interestingly, this came almost exclusively from healthcare workers who were also patients within the system).

In an attempt to mitigate the limitation of only getting data from consented research participants, the investigators approached Geisinger patients who declined to participate in MyCode for permission to discuss the reasons for their decision. Most of the patients indicated that their medical issues and/or life situation were not conducive to participation at the time of invitation. A few indicated concerns about privacy or discrimination. Very few noted that the potential return of medically relevant results led to their decision (unpublished data). These findings mirror those reported by the CSER consortium. These investigations are ongoing so that the issues leading to non-participation can be fully explored.

To date, over 1000 MyCode participants have received results from sequencing.[13] These participants were contacted 6 months after the result was returned to assess the outcomes related to the information (Williams et al., 2018). As with the other projects, the majority expressed no increase in anxiety, distress, or other concerns after learning of their results. Many expressed gratitude for the result. Some noted that the result explained personal or family medical history, which represents a reduction of scientific uncertainty. Other participants related initial distress at receiving the

---

[13] Retrieved June 6, 2018, from https://www.geisinger.org/mycode-results.

unexpected result, however by the 6-month time point, this had resolved, with many now expressing a positive response to the result based on an increased perception of control over the risk identified through sequencing, as an aspect of personal uncertainty. None of the participants receiving a result indicated that they had a significant adverse psychological reaction, nor utilized the psychological support services provided as part of the project. These findings are consistent with those reported from the other studies. This investigation is ongoing with publication of findings expected in the near future.

CONCLUSION

NGS is rapidly emerging into clinical practice for a variety of indications. While it is important to continue to elucidate the depth and breadth of potential psychological implications of learning genomic sequencing results—particularly in the setting of additional findings or population screening programs—the evidence, while still limited, does not support concern about undue psychological impact of these results. These questions are the subject of ongoing study and should continue to be a priority so that empiric evidence can be generated to inform the development of clinical guidance.

REFERENCES

Athens, B. A., Caldwell, S. L., Umstead, K. L., Connors, P. D., Brenna, E., & Biesecker, B. B. (2017). A systematic review of randomized controlled trials to assess outcomes of genetic counseling. *Journal of Genetic Counseling, 26,* 902–933.

Best, M., Newson, A. J., Meiser, B., Juraskova, I., Goldstein, D., Tucker, K., ... Butow, P. (2018). The PiGeOn project: Protocol of a longitudinal study examining psychosocial and ethical issues and outcomes in germline genomic sequencing for cancer. *BMC Cancer, 18,* 454.

Biesecker, L. G., Mullikin, J. C., Facio, F. M., Turner, C., Cherukuri, P. F., Blakesley, R. W., ... Young, A. C. (2009). The ClinSeq Project: Piloting large-scale genome sequencing for research in genomic medicine. *Genome Research, 19,* 1665–1674.

Bode, A. M., & Dong, Z. (2017). Precision oncology—The future of personalized cancer medicine? *NPJ Precision Oncology, 1,* 2.

Bonis, P. A., Trikalinos, T. A., Chung, M., Chew, P., Ip, S., DeVine, D. A., & Lau, J. (2007). Hereditary nonpolyposis colorectal cancer: Diagnostic strategies and their implications. *Evidence Report/Technology Assessment* (150), 1–180.

Carey, D. J., Fetterolf, S. N., Davis, F. D., Faucett, W. A., Kirchner, H. L., Mirshahi, U., ... Ledbetter, D. H. (2016). The Geisinger MyCode community health initiative: An electronic health record-linked biobank for precision medicine research. *Genetics in Medicine, 18,* 906–913.

Chae, Y. K., Pan, A. P., Davis, A. A., Patel, S. P., Carneiro, B. A., Kurzrock, R., & Giles, F. J. (2017). Path toward precision oncology: Review of targeted therapy studies and tools to aid in defining "actionability" of a molecular lesion and patient management support. *Molecular Cancer Therapeutics, 16,* 2645–2655.

Christensen, C. M., Grossman, J. H., & Hwang, J. (2009). *The Innovator's Prescription a disruptive solution for health care.* New York: McGraw-Hill.

Christensen, K. D., Roberts, J. S., Uhlmann, W. R., & Green, R. C. (2011). Changes to perceptions of the pros and cons of genetic susceptibility testing after APOE genotyping for Alzheimer disease risk. *Genetics in Medicine, 13*, 409−414.

Collins, F. S., & Varmus, H. (2015). A new initiative on precision medicine. *The New England Journal of Medicine, 372*, 793−795.

Córdoba, M., Rodriguez-Quiroga, S. A., Vega, P. A., Salinas, V., Perez-Maturo, J., Amartino, H., ... Kauffman, M. A. (2018). Whole exome sequencing in neurogenetic odysseys: An effective, cost- and time-saving diagnostic approach. *PLoS One, 13*(2), e0191228.

Ewans, L. J., Schofield, D., Shrestha, R., Zhu, Y., Gayevskiy, V., Ying, K., ... Roscioli, T. (2018). Whole-exome sequencing reanalysis at 12 months boosts diagnosis and is cost-effective when applied early in Mendelian disorders. *Genetics in Medicine.* Available from https://doi.org/10.1038/gim.2018.39, [Epub ahead of print].

Faucett, W. A., & Davis, F. D. (2016). How Geisinger made the case for an institutional duty to return genomic results to biobank participants. *Applied & Translational Genomics, 8*, 33−35.

Ferrer, R. A., Taber, J. M., Klein, W. M., Harris, P. R., Lewis, K. L., & Biesecker, L. G. (2015). The role of current affect, anticipated affect and spontaneous self-affirmation in decisions to receive self-threatening genetic risk information. *Cognition & Emotion, 29*, 1456−1465.

Genetic Testing Registry Search. Retrieved June 10, 2018, from https://www.ncbi.nlm.nih.gov/gtr/all/tests/?term = Human + genome%5BTESTTARGET%5D + OR + Whole + exome%5BTESTTARGET%5D.

Genomics and World Health. Report of the Advisory Committee on Health Research, Geneva, WHO (2002) WHA 57.13. Genomics and World Health, Fifty Seventh World Health Assembly Resolution; 22 May 2004.

Graves, K. D., Vegella, P., Poggi, E. A., Peshkin, B. N., Tong, A., Isaacs, C., ... Schwartz, M. D. (2012). Long-term psychosocial outcomes of BRCA1/BRCA2 testing: Differences across affected status and risk-reducing surgery choice. *Cancer Epidemiology, Biomarkers & Prevention, 21*, 445−455.

Gray, S. W., Hicks-Courant, K., Lathan, C. S., Garraway, L., Park, E. R., & Weeks, J. C. (2012). Attitudes of patients with cancer about personalized medicine and somatic genetic testing. *Journal of Oncology Practice, 8*, 329−335, 2 p following 335.

Haga, S. B., Mills, R., & Bosworth, H. (2014). Striking a balance in communicating pharmacogenetic test results: Promoting comprehension and minimizing adverse psychological and behavioral response. *Patient Education and Counseling, 97*, 10−15.

Han, P. K. J., Klein, W. M. P., & Arora, N. K. (2011). Varieties of uncertainty in health care. *Medical Decision Making, 31*, 828−838.

Haslem, D. S., Chakravarty, I., Fulde, G., Gilbert, H., Tudor, B. P., Lin, K., ... Nadauld, L. D. (2018). Precision oncology in advanced cancer patients improves overall survival with lower weekly health-care costs. *Oncotarget, 9*, 12316−12322.

Hosseini, S. M., Kim, R., Udupa, S., Costain, G., Jobling, R., Liston, E., ... Gollob, M. H., & NIH-Clinical Genome Resource Consortium. (2018). Reappraisal of reported genes for sudden arrhythmic death: An evidence-based evaluation of gene validity for Brugada syndrome. *Circulation, 138*, 1195−1205.

Jones, L. K., Kulchak Rahm, A., Gionfriddo, M. R., Williams, J. L., Fan, A. L., Pulk, R. A., ... Williams, M. S. (2018). Developing pharmacogenomic reports: Insights from patients and clinicians. *Clinical and Translational Science, 11*, 289−295.

Lin, X., Tang, W., Ahmad, S., Lu, J., Colby, C. C., Zhu, J., & Yu, Q. (2012). Applications of targeted gene capture and next-generation sequencing technologies in studies of human deafness and other genetic disabilities. *Hearing Research, 288*, 67−76.

Lumish, H. S., Steinfeld, H., Koval, C., Russo, D., Levinson, E., Wynn, J., ... Chung, W. K. (2017). Impact of panel gene testing for hereditary breast and ovarian cancer on patients. *Journal of Genetic Counseling, 26*, 1116−1129.

McConkie-Rosell, A., Hooper, S. R., Pena, L. D. M., Schoch, K., Spillmann, R. C., Jiang, Y. H., ... Shashi, V. (2018). Psychosocial profiles of parents of children with undiagnosed diseases: Managing well or just managing? *Journal of Genetic Counseling, 27*(4), 935−946. Available from https://doi.org/10.1007/s10897-017-0193-5, [Epub 2018 Jan 2. PubMed PMID: 29297108; PubMed Central PMCID: PMC6028305].

Meng, L., Pammi, M., Saronwala, A., Magoulas, P., Ghazi, A. R., Vetrini, F., ... Lalani, S. R. (2017). Use of exome sequencing for infants in intensive care units: Ascertainment of severe single-gene disorders and effect on medical management. *JAMA Pediatrics*, *171*, e173438.

Nakagawa, H., & Fujita, M. (2018). Whole genome sequencing analysis for cancer genomics and precision medicine. *Cancer Science*, *109*, 513—522.

Olson, J. E., Rohrer Vitek, C. R., Bell, E. J., McGree, M. E., Jacobson, D. J., St Sauver, J. L., ... Bielinski, S. J. (2017). Participant-perceived understanding and perspectives on pharmacogenomics: The Mayo Clinic RIGHT protocol (Right Drug, Right Dose, Right Time). *Genetics in Medicine*, *19*, 819—825.

Oonk, A. M. M., Ariens, S., Kunst, H. P. M., Admiraal, R. J. C., Kremer, H., & Pennings, R. J. E. (2018). Psychological impact of a genetic diagnosis on hearing impairment—An exploratory study. *Clinical Otolaryngology*, *43*, 47—54.

Pauker, S. G., & Kassirer, J. P. (1987). Decision analysis. *The New England Journal of Medicine*, *316*, 250—258.

Peay, H. L., Meiser, B., Kinnett, K., Furlong, P., Porter, K., & Tibben, A. (2016). Mothers' psychological adaptation to Duchenne/Becker muscular dystrophy. *European Journal of Human Genetics*, *24*, 633—637.

Picot, J., Bryant, J., Cooper, K., Clegg, A., Roderick, P., Rosenberg, W., & Patch, C. (2009). Psychosocial aspects of DNA testing for hereditary hemochromatosis in at-risk individuals: A systematic review. *Genetic Testing and Molecular Biomarkers*, *13*, 7—14.

Rahman, B., Meiser, B., Sachdev, P., Barlow-Stewart, K., Otlowski, M., Zilliacus, E., & Schofield, P. (2012). To know or not to know: An update of the literature on the psychological and behavioral impact of genetic testing for Alzheimer disease risk. *Genetic Testing and Molecular Biomarkers*, *16*, 935—942.

Sands, T. T., & Choi, H. (2017). Genetic testing in pediatric epilepsy. *Current Neurology and Neuroscience Reports*, *17*, 45.

Schallenberg, S., Merkelbach-Bruse, S., & Buettner, R. (2017). Lung cancer as a paradigm for precision oncology in solid tumours. *Virchows Archiv*, *471*, 221—233.

Schwarze, K., Buchanan, J., Taylor, J. C., & Wordsworth, S. (2018). Are whole-exome and whole-genome sequencing approaches cost-effective? A systematic review of the literature. *Genetics in Medicine*, *20*(10), 1122—1130. Available from https://doi.org/10.1038/gim.2017.247, [Epub 2018 Feb 15. Review. PubMed PMID: 29446766].

Sholl, L. (2017). Molecular diagnostics of lung cancer in the clinic. *Translational Lung Cancer Research*, *6*, 560—569.

Skinner, D., Raspberry, K. A., & King, M. (2016). The nuanced negative: Meanings of a negative diagnostic result in clinical exome sequencing. *Sociology of Health & Illness*, *38*, 1303—1317.

Skinner, D., Roche, M. I., Weck, K. E., Raspberry, K. A., Foreman, A. K. M., Strande, N. T., ... Henderson, G. E. (2018). "Possibly positive or certainly uncertain?": Participants' responses to uncertain diagnostic results from exome sequencing. *Genetics in Medicine*, *20*, 313—319.

Smith, L. D., Willig, L. K., & Kingsmore, S. F. (2015). Whole-exome sequencing and whole-genome sequencing in critically ill neonates suspected to have single-gene disorders. *Cold Spring Harbor Perspectives in Medicine*, *6*(2), a023168.

Stark, Z., Schofield, D., Martyn, M., Rynehart, L., Shrestha, R., Alam, K., ... White, S. M. (2018). Does genomic sequencing early in the diagnostic trajectory make a difference? A follow-up study of clinical outcomes and cost-effectiveness. *Genetics in Medicine*. Available from https://doi.org/10.1038/s41436-018-0006-8, [Epub ahead of print].

Taber, J. M., Klein, W. M., Ferrer, R. A., Han, P. K., Lewis, K. L., Biesecker, L. G., & Biesecker, B. B. (2015). Perceived ambiguity as a barrier to intentions to learn genome sequencing results. *Journal of Behavioral Medicine*, *38*, 715—726.

Taber, J. M., Klein, W. M., Ferrer, R. A., Lewis, K. L., Biesecker, L. G., & Biesecker, B. B. (2015). Dispositional optimism and perceived risk interact to predict intentions to learn genome sequencing results. *Health Psychology*, *34*, 718—728.

Taber, J. M., Klein, W. M., Ferrer, R. A., Lewis, K. L., Harris, P. R., Shepperd, J. A., & Biesecker, L. G. (2015). Information avoidance tendencies, threat management resources, and interest in genetic sequencing feedback. *Annals of Behavioral Medicine*, *49*, 616—621.

Tzeng, J. P., Mayer, D., Richman, A. R., Lipkus, I., Han, P. K., Valle, C. G., . . . Brewer, N. T. (2010). Women's experiences with genomic testing for breast cancer recurrence risk. *Cancer, 116,* 1992–2000.

Vissers, L. E. L. M., van Nimwegen, K. J. M., Schieving, J. H., Kamsteeg, E. J., Kleefstra, T., Yntema, H. G., . . . Willemsen, M. A. A. P. (2017). A clinical utility study of exome sequencing versus conventional genetic testing in pediatric neurology. *Genetics in Medicine, 19,* 1055–1063.

Werner-Lin, A., Zaspel, L., Carlson, M., Mueller, R., Walser, S. A., Desai, R., & Bernhardt, B. A. (2018). Gratitude, protective buffering, and cognitive dissonance: How families respond to pediatric whole exome sequencing in the absence of actionable results. *American Journal of Medical Genetics. Part A, 176,* 578–588.

Williams, M. S., Buchanan, A. H., Davis, F. D., Faucett, W. A., Hallquist, M. L. G., Leader, J. B., . . . Ledbetter, D. H. (2018). Patient-centered precision health in a learning health care system: Geisinger's genomic medicine experience. *Health Affairs, 37,* 757–764.

Wynn, J., Holland, D. T., Duong, J., Ahimaz, P., & Chung, W. K. (2018). Examining the psychosocial impact of genetic testing for cardiomyopathies. *Journal of Genetic Counseling, 27*(4), 927–934. Available from https://doi.org/10.1007/s10897-017-0186-4, [Epub 2017 Dec 15. PubMed PMID: 29243008].

FURTHER READING

Nadauld, L. D., Ford, J. M., Pritchard, D., & Brown, T. (2018). Strategies for clinical implementation: Precision oncology at three distinct institutions. *Health Affairs, 37*(5), 751–756.

CHAPTER 8

Genetic Counseling and Genomic Sequencing

Barbara A. Bernhardt

University of Pennsylvania, Philadelphia, PA, United States

INTRODUCTION

Traditionally, genetic counseling has focused on helping patients and families understand and adapt to the contribution of genetics to disease (Resta et al., 2006). As important outcomes of having received genetic counseling, patients should be empowered to make informed decisions, make effective use of available medical and support services, manage their feelings, and look toward the future with hope (McAllister, Wood, Dunn, Shiloh, & Todd, 2011). Generally, patient decisions have revolved around whether or not to undergo genetic testing.

For decades, referrals for genetic counseling originated because of a personal or family history of a suspected genetic condition. A genetic counselor would then offer appropriate genetic testing to identify potentially causative mutations in one or several suspect genes. The type of result available was predictable, and most of the results given to patients were interpretable. Patients (or family members) would often be familiar with the condition for which testing is offered (Ormond, 2013), may have given thought to whether or not to be tested, and have expectations about their responses to positive results (Biesecker, 2016a). The genetic counselor would support informed decisions by providing in-depth education about genetic principles, how testing is performed, the types and implications of possible results, and the risks, benefits, and limitations of testing.

With technological advances, the menu and technical complexity of tests available to patients has grown and continues to expand. Instead of testing for genes associated with a particular disease as previously performed, genomic testing can identify variants throughout the genome, including variants that might be associated with a patient's presenting symptoms, as well as additional secondary or incidental findings. These might include variants in genes for other Mendelian conditions, pharmacogenomic results, carrier status, and variants associated with an increased risk for common complex conditions. The number and scope of results available through genomic

Clinical Genome Sequencing
DOI: https://doi.org/10.1016/B978-0-12-813335-4.00008-8

sequencing is potentially large, and the clinical implications of some detected variants remain unknown, even after interpreting findings through the lens of the patient's medical and family history (Han et al., 2017). Despite the limitations of genomic testing, extensive media attention and hype have heightened public expectations about the type and utility of information that can be revealed and acted upon (Caulfield, 2016; O'Rourke, 2013). The need to temper patient expectations about what can be gained from sequencing, especially in the face of universally low genetic literacy, further complicates informed decision-making, but will be necessary in order to avoid poor adjustment to findings (Anderson et al., 2017; Bernhardt et al., 2015; Bowdin et al., 2016; Hooker, Ormond, Sweet, & Biesecker, 2014).

Given the complexity of genomic sequencing, and the number, range, and uncertainty of possible results, providing in-depth education about technological aspects of testing and about all potential results in pretest genetic counseling will be impossible (Austin, Semaka, & Hadjipavlou, 2014; Merrill & Guthrie, 2015). Yet, patients need to make decisions about whether to undergo testing, and about learning various types of incidental findings that might be available to be returned. Such decision-making may be complicated because patients offered genomic sequencing, both those with a condition and those who are healthy, may not have familiarity with the types of conditions that may be detected as secondary findings, and lack a "lived experience" with the conditions (Ormond, 2013).

Leaders in the field have called on the profession to carefully examine extant approaches to genetic counseling to make careful and deliberate adaptations to enable sound pre- and post-test "genomic counseling" (Austin et al., 2014; Biesecker, 2016b; Merrill & Guthrie, 2015; Ormond et al., 2010). Such evaluations must include identifying evidence-based benchmarks to ensure that patients are making informed decisions, have realistic expectations of findings they might receive, understand the findings they do receive, and manage their results appropriately (Yu et al., under review). The challenges of genomic counseling and the processes needed to address these will be discussed in this chapter.

WHAT DO PATIENTS EXPECT FROM GENOMIC SEQUENCING?

Expectations from genomic sequencing will vary according to the indication for testing. Generally, genomic sequencing is viewed as a comprehensive test that may provide an answer to an existing health problem as well as information that might predict future disease (Halverson, Clift, & McCormick, 2016). Frequently, a child is being tested because of a suspected genetic condition. Such a child may have had previous genetic testing with negative results, and the parents may be on a long (and mostly discouraging) "diagnostic odyssey" seeking answers for their child (Hayeems, Babul-Hirji, Hoang, Weksberg, & Shuman, 2016; Watson, Hayes, & Radford-Paz, 2011).

The parents may view genomic sequencing as their final hope, and thus have high expectations about finally getting an answer (Krabbenborg et al., 2016). Additionally, parents frequently expect that when the cause of a condition is identified, there will be increased clarity about the prognosis and how that condition should be managed (Watson et al., 2011).

When sequencing is offered to oncology patients, the novelty of the testing, and the perception that target treatments might follow leads some patient to believe that genomic information holds special promise (Miller et al., 2014). When testing is done to guide *future* treatment, as is the case for many oncology patients, there may be the erroneous expectation that sequencing results can be used for *immediate* treatment (Malek et al., 2017).

For healthy adults seeking genomic sequencing, expectations to receive useful health-related information for the person tested, as well as for family members, are common (Facio et al., 2011; Haase, Michie, & Skinner, 2015; Linderman, Nielsen, & Green, 2016; Robinson et al., 2016; Sanderson et al., 2016). Patients may also expect to receive information about ancestry (Facio et al., 2011). Some being sequenced report being uncertain about the types of results they might receive, and how those results might be useful (Hylind, Smith, Rasmussen-Torvik, & Aufox, 2017).

Patients and families often overestimate the quantity, utility, and certainty of information that can be provided through genomic sequencing (Caulfield, 2016; Linderman et al., 2016; O'Rourke, 2013; Roche & Berg, 2015). Such unrealistic expectations are likely fueled by clinicians' presentation of the test as the most comprehensive diagnostic test available, and overly optimistic information provided by genomic testing websites and media portrayals (Roche & Berg, 2015; Skinner, Raspberry, & King, 2016; Walser et al., 2017). Misperceptions, including the belief that a negative result excludes a genetic diagnosis, the expectation that sequencing will find multiple actionable secondary or incidental findings relevant to future health, and that the test will provide a definitive diagnosis, are common (Bernhardt et al., 2015; Linderman et al., 2016; Malek et al., 2017). Failure of genomic sequencing to provide clear results with clinical utility often leads to patient disappointment and frustration.

Compounding these inflated expectations, the notion that information in and of itself has utility is common (Halverson et al., 2016; Malek et al., 2017). The idea that "knowledge is power" may lead to a perceived moral obligation to learn all possible information about health risks for both the individual tested and for family members (Malek et al., 2017; Miller et al., 2014). For example, parents, if given the option, generally will opt to learn all potential results from sequencing, especially if they anticipate future regret from not having made such a decision (Bishop, Strong, & Dimmock, 2017; Cornelis et al., 2016; Levenseller et al., 2014; Roche & Berg, 2015).

PROVIDING GENOMIC COUNSELING

Balancing supportive and informational needs

Genetic counselors have embraced the educational goals of genetic counseling. The majority of genetic counselors adopt the teaching-oriented model of genetic counseling, structuring sessions around the provision of biomedical information to patients, while leaving psychosocial issues largely unaddressed (Meiser, Irle, Lobb, & Barlow-Stewart, 2008; Paul, Metcalfe, Stirling, Wilson, & Hodgson, 2015; Roter, Ellington, Erby, Larson, & Dudley, 2006). As research reveals more about the human genome and as genomic testing becomes more complex, the amount of biomedical information that could be included in genomic counseling sessions is huge. Many clinicians may respond to this growing complexity by increasing the amount of information provided to patients and families, leaving less time to address the equally increasingly complex psychosocial issues and concerns that patients bring to sessions (Samuel, Dheensa, Farsides, Fenwick, & Lucassen, 2017). Since addressing emotional issues is associated with improved acquisition of knowledge (Austin et al., 2014; Veach, Bartels, & LeRoy, 2007), adopting an approach that emphasizes the "counseling" part of genomic counseling is urgently needed (Austin et al., 2014; Samuel et al., 2017). Traditional psychosocial genetic counseling techniques including contracting with patients to create a mutual agenda for both pre- and post-test counseling sessions, building a relationship of trust, addressing verbal and nonverbal cues, and directly acknowledging and validating the patients' emotions and experiences will strengthen the patient–provider relationship (Werner-Lin, McCoyd, & Bernhardt, 2016) leading to improved processing of complex biomedical information (Austin et al., 2014; Meiser et al., 2008; Veach et al., 2007) and greater satisfaction with the testing experience overall.

The Reciprocal Engagement Model of genetic counseling highlights the use of psychosocial approaches to the provision of information (Veach et al., 2007). Within this model, instead of providing a recitation of information the counselor thinks the patient should know, the counselor helps patients to determine the type of information which will be important to them in decision-making (Hooker et al., 2014). Furthermore, the counselor evaluates what the patient already understands and how s/he would like to learn new information to so as to personalize the delivery of information in the context of prior knowledge and preferred learning style (Biesecker, 2016b).

Central to the Reciprocal Engagement Model is the counselor–patient relationship. This relationship is built on empathic listening, inquiring, and positive regard. Trust is increased when empathic counselors reflect emotions back to patients so they know they are understood (Biesecker, 2016b). Such counseling techniques are likely to resonate with genomic counseling patients who, as Rosell et al. (2016) have shown, desire trusting relationships based on respect, understanding, and open communication. Furthermore, acknowledging and empathically responding to emotional cues exhibited by the patient, including negative emotions such as frustration, anger, and disappointment, creates a "safe" environment in which patients feel free to share their emotions, further solidifying the patient—clinician relationship (Werner-Lin, McCoyd, & Bernhardt, 2016).

Personalizing personalized medicine

The broad scope and uncertainty regarding potential results, in combination with patients' unrealistic expectations about the quantity and utility of results, remains a pressing challenge for pretest counseling. A focus on providing information to address the complexity of sequencing and all possible results is inadvisable given that genetic counseling patients generally report having received too much information that is excessively complex, and not enough information that has personal relevance (Joseph et al., 2017; Merrill & Guthrie, 2015; Yu et al., under review). Such an effort to personalize pretest counseling may begin with the clinician exploring how the patient expects to benefit from genomic sequencing. Families may expect benefits beyond the clinical benefits highlighted by researchers and clinicians (Anderson et al., 2017; Cacioppo, Chandler, Towne, Beggs, & Holm, 2016; Halverson et al., 2016; Hylind et al., 2017; Kohler, Turbitt, & Biesecker, 2017; Malek et al., 2017; Rosell et al., 2016). For example, parents of children with serious medical conditions expect to accrue benefits such as peace of mind, relief from guilt, or information for reproductive planning (Malek et al., 2017). In addition to uncovering family-specific expected benefits of sequencing, such discussions can identify misperceptions about potential benefits, and redirect families to realistic expectations.

Pretest counseling provides an opportunity for the counselor to review the clinical and psychosocial implications of various types of results (positive, negative, uncertain) relating to the patient's presenting condition, as well as their anticipated emotional response to each type of result. This time also permits the counselor to explicitly address, and prepare the patient for, potential uncertainties of test results to calibrate their expectations and reduce disappointment if results with uncertain implications are returned (Biesecker et al., 2017).

Counselors may then address the broad categories of secondary findings that might be available to be returned (this may be laboratory- or institution-specific) and help the patient decide which, if any, of these results might be desired. Because patients frequently believe that secondary results will shed light on their risk for conditions present in the family (Werner-Lin, Tomlinson, Miller, & Bernhardt, 2016), explicitly asking the patient about results they expect or hope to learn is suggested. Counselors can tailor this discussion by helping the patient articulate those topics they wish to address to facilitate decision-making. Other topics may be mentioned only briefly or held for future discussion (Yu et al., under review).

Such an approach is likely to result in providing a minimal amount of information about genes and genomic sequencing technology. This is sufficient in pretest counseling since these topics are often unnecessary for informed decision-making (Yu et al., under review). To be "informed," patients and families must have command of the possibilities, uncertainties, and consequences of their testing decisions rather than the intricacies of how testing is performed or how genes function. Such discussions likely will help the patient to develop realistic expectations about the various possibilities, uncertainties, and implications of results, including secondary findings (Bernhardt et al., 2015).

Special considerations

The context and timing of genomic counseling in the patient's life course merit the counselor's attention. Patients may approach genetic counselors at varying points in the course of a disease; each phase incurs unique psychosocial challenges and information needs (Rolland & Werner-Lin, 2006). Further, loved ones may or may not have been diagnosed with or died of a familial condition, and grief may complicate patient reactions to approaching counseling, considering testing, and understanding results. The genetic counselor should identify these facets of the patient's experience and modify sessions accordingly.

If the patient offered sequencing has a potentially life-limiting condition, such as a metastatic cancer, the patient or the surrogate decision-maker may feel unprepared to make an informed decision about undergoing sequencing or learning secondary findings, and the option of deferring or declining testing can be discussed (Druker et al., 2017; Scollon et al., 2014). In such cases, counselors may invite consideration of banking DNA so that testing could be done in the future to benefit other family members (Raymond et al., 2016). Clinicians might also want to ascertain and document the patient's wishes regarding to whom results should be disclosed in the event the patient is unable to receive them, for example, if the patient is too ill or is deceased (Raymond et al., 2016).

With the advent of genomic sequencing, testing will increasingly be offered to children, including adolescents (Biesecker, 2016a). Until recently, because of concerns about preserving the child's autonomy and the potential for adverse psychosocial responses, nondiagnostic genetic testing of children was rarely performed (Ross, Saal, David, Anderson, & American Academy of Pediatrics, 2013; Wakefield et al., 2016). As a consequence, there is limited experience with offering genomic testing or test results to children and adolescents, and few data available on their response to testing (Wakefield et al., 2016). Although the involvement of older children and adolescents in pretest counseling and decision-making is encouraged (Botkin et al., 2015), most children are unfamiliar with genetic counseling (Pichini et al., 2016), and unaccustomed to being involved in clinical decision-making (Werner-Lin, Tomlinson, et al., 2016).

Attempts to involve adolescents in pretest genomic counseling sessions by assessing genetics knowledge generally do not lead to engagement (Werner-Lin, Tomlinson, et al., 2016). Rather, engagement can be activated by directing parts of the conversation to them, providing guidance about the kinds of questions they might ask, giving the child explicit permission to express their own opinions, repeatedly inviting them to engage in dialogue and ask questions, and checking in for understanding throughout the session (Miller, Werner-Lin, Walser, Biswas, & Bernhardt, 2017; Pichini et al., 2016; Werner-Lin, Tomlinson, et al., 2016). These behaviors show respect for the adolescent and place him or her in the center of decision-making (Miller et al., 2017). Since sequencing has implications for family groups, pretest counseling sessions with adolescents and their parents may need to focus on the family's decision-making process as well as individual family members' motivations and expectations (Werner-Lin, Tomlinson, et al., 2016). Counselors may evaluate how families have made medical decisions, or other important decisions regarding the adolescent, in the past, invite the family to consider the process of these decisions, and make allowances for the adolescent's developing capacity for autonomy and self-determination.

Before concluding the pretest genomic counseling session with an adolescent patient, the clinician and family should discuss how results will be returned to the family, including whether results will be returned to the child and the parents together, or to the parents first, and then to the adolescent.

RETURNING RESULTS FROM GENOMIC SEQUENCING
How do families respond to results from genomic sequencing?

To date there are limited data about how families respond to genomic sequencing results. Some preliminary observations have been reported from the Clinical Sequencing Exploratory Research Consortium (Amendola et al., 2015; Skinner et al., 2016, 2018; Wynn et al., 2018), and from other clinical or research projects (Anderson et al., 2017; Halverson et al., 2016; Linderman et al., 2016; Rosell et al., 2016). Not surprisingly, the response of research participants and patients to results is variable, and depends on factors such as the types and clarity of results returned, the indication for testing, previous experience with genetic testing, and patient expectations for results (Wynn et al., 2018). Responses are also shaped by factors such as individual and family disease histories, access to healthcare and support services, and individual psychological attributes. The same finding may provoke a range of feelings from patient to patient, in members of the same family, and in distinct communities. Individuals may also respond with a range of intensities in their emotions, which may change over time. Counselors must therefore be prepared to address a variety of emotions, of varying intensities, over the course of one or more visits.

When receiving positive diagnostic results, many families are grateful to have an explanation for the patient's condition, and anticipate a reduction in uncertainty about etiology, prognosis, recommended management, and recurrence risk. Families who feared that sequencing would yield a diagnosis that is associated with early death may be relieved when they learn that the diagnosed condition is less severe than anticipated (Rosell et al., 2016). Ending the diagnostic odyssey, though, frequently leads families on yet another odyssey as they search for information about the newly diagnosed condition, establish a new care plan, meet with new providers, seek out specialists, and search for new peer contacts (Krabbenborg et al., 2016).

Family responses are tempered by the nature of the condition diagnosed. Shock, anticipatory loss, and grief are common when the results point toward a progressive disorder, one that is difficult to treat, or one that is more serious than anticipated (Marron et al., 2016; Wynn et al., 2018). Moreover, when the diagnosed condition is rare, patients and family members are often distressed to learn that limited or no information exists to guide medical management, or that they have no access to appropriate peer or patient support groups (Cacioppo et al., 2016; Krabbenborg et al., 2016).

When diagnostic results are negative or uncertain, responses are also mixed, ranging from disappointment to relief (Wynn et al., 2018). Some families are relieved because nothing serious was found, or because they now believe that the condition is not genetic, meaning other family members and/or future children are not at increased risk (Krabbenborg et al., 2016; Wynn et al., 2018). When patients and families had high expectations that genomic sequencing would find an explanation for a child's or their own condition, they were often disappointed by the absence of diagnostic findings. Providers may reassure patients, or instill hope in the face of negative or uncertain results that a future diagnosis is still possible as researchers learn more about rare variants, or as sequencing technology improves (Skinner et al., 2016). Patients are generally pleased to be told that they are contributing personally to scientific advances in understanding the implications of uncertain variants (Skinner et al., 2018).

In some situations where results are uncertain, or when biomedical information is unavailable or unclear, patients may seek other sources of knowledge beyond the clinician in order to reduce uncertainty, including possibly searching for families with a similar uncertain variant. (Halverson et al., 2016; Rubel, Werner-Lin, Barg, & Bernhardt, 2017). Families also might personally seek information about variant reclassification through medical experts and public databases, as well as through other sources, such as the FindMyVariant project through the University of Washington.

As sequencing was being introduced through research protocols and in clinical care, many voiced concerns that secondary results might negatively affect patients by revealing unwanted, overwhelming, or uncertain information, or be misunderstood and lead to inappropriate medical care (Biesecker, Burke, Kohane, Plon, & Zimmern, 2012; Burke et al., 2013; Clarke, 2014; Klitzman, Appelbaum, & Chung, 2013).

In practice, responses to secondary results, or to the absence of them, are complex. Patients are often pleased when they receive results that can be acted upon medically or behaviorally so as to reduce risk. Other patients are surprised and anxious about secondary results, especially when a result suggests a risk for a condition that was not anticipated based on the family history, or one thought to be medically serious (Wynn et al., 2018). When there are no secondary findings, some patients feel relief, believing they are not at increased risk for future disease (Walser et al., 2017). Patients who expected to receive secondary findings predicting risks for conditions that could be medically managed may be disappointed when there are no secondary results to guide future medical care (Linderman et al., 2016; Wynn et al., 2018), and some believe that results may have been withheld (Werner-Lin et al., 2018).

Because of the limited experience offering genomic testing to children, few data are available addressing the long-term implications of pediatric testing on children's medical care, psychosocial functioning, or relationships. The information that is available from testing children for single-gene conditions suggests that there is minimal evidence of adverse short-term outcomes (Wakefield et al., 2016). However, data are lacking that explore long-term outcomes, potential disruptions to children's developing identities, family dynamics, parent–child relationships, or survivor guilt (Wakefield et al., 2016).

How should results be discussed with patients and families?

The disclosure of results from genomic sequencing is challenging due to the potentially large number and uncertainty of results, patient's unrealistic expectations for useful information, the possible presence of unexpected secondary findings, and difficulty in predicting how patients and families will respond (Wynn et al., 2018). Some guiding principles are outlined that can assist clinicians as they discuss genomic sequencing results with patients and family members.

Information is necessary, but not sufficient

Frequently, clinicians are invested in providing clear explanations and information at the expense of monitoring patients' emotional reactions to the information. Since many patients have a limited capacity to process new information or multiple bits of information at one time, particularly when they are in heightened emotional states, clinicians will need to resist attempting to educate families about all aspects of all results in one session (Walser et al., 2017). Many patients experience powerful emotions when learning new genetic information, and report the experience to be overwhelming (Ashtiani, Makela, Carrion, & Austin, 2014). This can interfere with processing information, formulating questions, and engagement in sessions (Wynn et al., 2018). Indeed, when receiving genetic diagnoses, most patients and parents report being passive recipients of information without acknowledgment of emotions or direction about what to do next (Ashtiani et al., 2014). Clinicians should carefully monitor and respond to the patient and family emotional responses to results. Because many families have complex and rapidly shifting responses to a result, clinicians should acknowledge those emotions as well as implicit and explicit ambivalence (Werner-Lin, McCoyd, et al., 2016). When the patient's emotions are acknowledged, s/he will feel understood and prepared to listen to the information, and ready to act on it (Djurdjinovic, 2011).

Be flexible to permit families time to digest

Inserting frequent pauses during sessions, or possibly giving families time alone after delivering results so they can process information and formulate questions is recommended. Families can also be given the option to have results returned in two sessions, with primary findings delivered first and secondary findings delivered separately (Bowdin et al., 2016; Ormond & Cho, 2014). Others have advocated for a delivery model in which results are delivered in stages at the appropriate time over the course of a lifetime (Mollison & Berg, 2017).

Target explanations to family literacy and interests

With so much information to discuss, clinicians may be unsure how to start or where to focus education. Some clinicians may structure the return of results sessions by sharing with the patient the laboratory report that summarizes findings, and going through the report and discussing each finding (Walser et al., 2017; Wynn et al., 2018). Since these reports are frequently lengthy, written at a high reading level, and loaded with unfamiliar scientific terms (Vassy et al., 2015; Walser et al., 2017), such structuring may result in sessions that are characterized by less patient engagement, reduced patient satisfaction, poor understanding of results, heavy use of jargon, and discussion of genomics that place a high literacy burden on families (Roter, Erby, Larson, & Ellington, 2007; Schnitzler et al., 2017; Walser et al., 2017). Instead, briefly summarizing the findings, and then taking guidance from the patient about which findings they would like to discuss in more detail is likely to result in sessions that are more understandable and responsive to patient needs (Schmidlen et al., 2018).

Although families vary in which aspects of genomic results they are most interested in discussing, many report a primary interest in the impact sequencing results will have on clinical care and health in general (Krabbenborg et al., 2016). Empiric research shows that in return of result sessions, families are most engaged when clinical implications are being discussed (Walser et al., 2017). Such discussions should address how (and if) results change medical management, as well as recommendations for additional testing and/or consultations. When results do not change management recommendations, families appreciate being told that they have been doing everything they could to manage their situation (Walser et al., 2017).

Discuss implications of results for family members

Genome sequencing results may imply risk for diseases and carrier status in other family members, thus potentially raising a host of other questions about the next steps for testing and medical follow-up. Families may find it difficult to imagine how they might discuss results with family members and others (Wynn et al., 2018). Clinicians may need to prepare patients for sharing results with family members, and can facilitate these discussions in a number of ways. First, help patients identify who would benefit from learning the information. Second, identify a time and place to tell these individuals about the results. Third, discuss with the patient what they will tell relatives, and how they anticipate family members will respond to the discussion.

Consider which clinicians or team members should be present when results are returned

When a diagnostic result is available, families want information provided by a knowledgeable provider (Watson et al., 2011). If the diagnostic or management issues relating to the results being returned are outside the expertise of the clinician who is returning the results, additional medical specialists may be invited to attend the session. Alternatively, the patient can receive a prompt referral to a specialist for expert information and guidance (Bowdin et al., 2016). When a number of providers return results together, each with their own focus, expertise, and interest in the case, the psychosocial issues and concerns of the family may be lost. Families may be intimidated by the number of experts in the room, the language used to describe the results, or the dynamics among providers. In this setting, genetic counselors ideally could act as translators and patient advocates, pacing sessions to meet family needs, addressing anxiety or worry, and verifying that the family is following and understanding the discussion.

Follow-up with families to support, clarify, and refer

Follow-up, either in person or by phone, is recommended to reinforce information, evaluate the need for psychologic support, discuss progress with recommended medical follow-up and sharing results with family members, and address additional questions and concerns. If a variant of uncertain significance was present, any expectations for reclassification should be discussed given the observation that families frequently expect that, if sequenced, they will receive continuous updates on the implications of results (Skinner et al., 2016, 2018).

After the consultation, most families seek out additional information to help them understand results or to seek out appropriate services. Identifying the ways patients gather information outside of the clinic is an important piece of the on-going assessment (Werner-Lin, McCoyd, et al., 2016). Clinicians can assist patients by advising them on how to search for and sort through the array of available online information

(Bernhardt et al., 2013; Walser, Werner-Lin, Russell, Wapner, & Bernhardt, 2016), including medical literature, links to support groups (online and in-person), and reputable websites (Rubel et al., 2017).

Special issues with pediatric sequencing

When a young child is tested, parents should be engaged in a discussion of when and how results will be shared with their child and his or her siblings. If the person sequenced is an adolescent, discussing the results with the parents and the child together, and offering opportunities to discuss results separately, is recommended (Duncan & Young, 2013; Werner-Lin et al., 2018). Discussion of results, their implications, and follow-up will need to be tailored to the individual child's developmental, cognitive, and emotional status. Children will have varying comfort and experiences in healthcare settings, and some will actively seek out information while others may appear to be disinterested. Some children may perseverate on future possibilities or negative emotions (Werner-Lin, Merrill, & Brandt, 2018). Clinicians will need to monitor the child during discussions for any evidence of distress or anxiety and address these responses with the child and the parents (Werner-Lin et al., 2018).

Children will need follow-up at various points to provide additional education, evaluate psychosocial needs, and to provide age-specific discussions of the implications of genomic results (Patenaude & Schneider, 2017). Follow-up genetic counseling in mid-to-late teenage years and for family planning can ensure that patients understand their diagnosis, risks, management recommendations, and reproductive risks and options (Druker et al., 2017).

CURRENT NEEDS AND FUTURE TRENDS

As genomic testing becomes more common and is increasingly offered by nongeneticists, clinicians who are unfamiliar with genomic testing will be charged with ordering tests and communicating results to patients (Borry et al., 2017). Broad education about genetics, genetic and genomic testing, and the genetic basis of disease is needed for clinicians and the public (Borry et al., 2017; Bowdin et al., 2016; Dougherty, Lontok, Donigan, & McInerney, 2014). To promote the successful integration of genomics into clinical settings, brief genomic testing reports, clinician decision support tools, and point of care educational strategies are under development (Christensen et al., 2018; Manolio et al., 2013; Pennington et al., 2017; Vassy et al., 2015). To complement face-to-face counseling, patient decision aids, simplified genomic test reports, and novel methods of delivering genomic test results directly to patients are being developed and evaluated (Birch, 2015; Brothers et al., 2017; Ekstract, Holtzman, Kim, Willis, & Zallen, 2017; Haga et al., 2014; Schmidlen et al., 2018; Tabor et al., 2017).

Since genetic counselors will factor prominently in service provision, strategies are needed for teaching both genetic counseling students and practicing genetic counselors to assist patients to effectively engage both cognitively and emotionally with genomic information (Hooker et al., 2014). Importantly, genetic counselors will need to be able to personally tolerate uncertainty, and develop skills to assist patients to manage uncertainty effectively (Han et al., 2017; Werner-Lin, McCoyd, et al., 2016). As genomic testing uncovers risks for medically actionable conditions, genetic counselors will be increasingly involved in health education and health promotion counseling. This will require that genetic counselors are trained in the underpinnings of health behavior change and health promotion models (Ormond, 2013).

To generate data for recommendations for best practices, validated measures of agreed-upon outcomes, both knowledge-based and psychosocial, of genomic counseling are needed (Phillips et al., 2017). Such measures could assess, for example, outcomes such as adaptation to risk, familial communication, rates of cascade testing, and screening and health-promotion behaviors (Wicklund & Trepanier, 2014). As measures become available, research is needed exploring the outcomes of various forms of pre- and post-test genomic counseling (Austin et al., 2014).

Advances in genomic technology and increased understanding of the contribution of genomic variants to health, disease, and drug response will result in the expansion of results reported from a single test, including information relevant to risks of diseases presenting from infancy to adulthood (Collins & Varmus, 2015). Strategies will be needed to effectively educate and support patients as they learn and utilize genomic information to improve health.

REFERENCES

Amendola, L. M., Lautenbach, D., Scollon, S., Bernhardt, B., Biswas, S., East, K., ... Wynn, J. (2015). Illustrative case studies in the return of exome and genome sequencing results. *Personalized Medicine*, *12*(3), 283–295.

Anderson, J. A., Meyn, M. S., Shuman, C., Shaul, R. Z., Mantella, L. E., Szego, M. J., ... Hayeems, R. Z. (2017). Parents perspectives on whole genome sequencing for their children: Qualified enthusiasm? *Journal of Medical Ethics*, *43*(8), 535–539.

Ashtiani, S., Makela, N., Carrion, P., & Austin, J. (2014). Parents' experiences of receiving their child's genetic diagnosis: A qualitative study to inform clinical genetics practice. *American Journal of Medical Genetics*, *164A*(6), 1496–1502.

Austin, J., Semaka, A., & Hadjipavlou, G. (2014). Conceptualizing genetic counseling as psychotherapy in the era of genomic medicine. *Journal of Genetic Counseling*, *23*(6), 903–909.

Bernhardt, B. A., Roche, M. I., Perry, D. L., Scollon, S. R., Tomlinson, A. N., & Skinner, D. (2015). Experiences with obtaining informed consent for genomic sequencing. *American Journal of Medical Genetics Part A*, *167*(11), 2635–2646.

Bernhardt, B. A., Soucier, D., Hanson, K., Savage, M. S., Jackson, L., & Wapner, R. J. (2013). Women's experiences receiving abnormal prenatal chromosomal microarray testing results. *Genetics in Medicine*, *15*(2), 139.

Biesecker, B. B. (2016a). Predictive genetic testing of minors: Evidence and experience with families. *Genetics in Medicine*, *18*(8), 763–764.

Biesecker, B. B. (2016b). The greatest priority for genetic counseling: Effectively meeting our clients' needs 2014 NSGC Natalie Weissberger Paul National Achievement Award. *Journal of Genetic Counseling, 25*(4), 621–624.

Biesecker, B. B., Woolford, S. W., Klein, W. M., Brothers, K. B., Umstead, K. L., Lewis, K. L., ... Han, P. K. (2017). PUGS: A novel scale to assess perceptions of uncertainties in genome sequencing. *Clinical Genetics, 92*(2), 172–179.

Biesecker, L. G., Burke, W., Kohane, I., Plon, S. E., & Zimmern, R. (2012). Next-generation sequencing in the clinic: Are we ready? *Nature Reviews Genetics, 13*(11), 818–824.

Birch, P. H. (2015). Interactive e-counselling for genetics pre-test decisions: Where are we now? *Clinical Genetics, 87*(3), 209–217.

Bishop, C. L., Strong, K. A., & Dimmock, D. P. (2017). Choices of incidental findings of individuals undergoing genome wide sequencing, a single center's experience. *Clinical Genetics, 91*(1), 137–140.

Borry, P., Bentzen, H. B., Budin-Ljøsne, I., Cornel, M. C., Howard, H. C., Feeney, O., ... Riso, B. (2017). The challenges of the expanded availability of genomic information: An agenda-setting paper. *Journal of Community Genetics*, 1–4.

Botkin, J. R., Belmont, J. W., Berg, J. S., Berkman, B. E., Bombard, Y., Holm, I. A., ... Wilfond, B. S. (2015). Points to consider: Ethical, legal, and psychosocial implications of genetic testing in children and adolescents. *The American Journal of Human Genetics, 97*(1), 6–21.

Bowdin, S., Gilbert, A., Bedoukian, E., Carew, C., Adam, M. P., Belmont, J., ... D'Alessandro, L. C. (2016). Recommendations for the integration of genomics into clinical practice. *Genetics in Medicine, 18*(11), 1075–1084.

Brothers, K. B., East, K. M., Kelley, W. V., Wright, M. F., Westbrook, M. J., Rich, C. A., ... Myers, J. A. (2017). Eliciting preferences on secondary findings: The Preferences Instrument for Genomic Secondary Results. *Genetics in Medicine, 19*(3), 337.

Burke, W., Antommaria, A. H., Bennett, R., Botkin, J., Clayton, E. W., Henderson, G. E., ... Press, N. A. (2013). Recommendations for returning genomic incidental findings? We need to talk!. *Genetics in Medicine, 15*(11), 854–859.

Cacioppo, C. N., Chandler, A. E., Towne, M. C., Beggs, A. H., & Holm, I. A. (2016). Expectation versus reality: The impact of utility on emotional outcomes after returning individualized genetic research results in pediatric rare disease research, a qualitative interview study. *PloS One, 11*(4), e0153597.

Caulfield, T. (2016). Ethics hype? *Hastings Center Report, 46*(5), 13–16.

Christensen, K. D., Bernhardt, B. A., Jarvik, G. P., Hindorff, L. A., Ou, J., Biswas, S., ... Goddard, K. A. B. (2018). Anticipated responses of early adopter genetic specialists and non-genetic specialists to unsolicited genomic secondary findings. *Genetics in Medicine, 20*(10), 1186–1195.

Clarke, A. J. (2014). Managing the ethical challenges of next-generation sequencing in genomic medicine. *British Medical Bulletin, 111*(1), 17–30.

Collins, F. S., & Varmus, H. (2015). A new initiative on precision medicine. *New England Journal of Medicine, 372*(9), 793–795.

Cornelis, C., Tibben, A., Dondorp, W., Van Haelst, M., Bredenoord, A. L., Knoers, N., ... Van Summeren, M. (2016). Whole-exome sequencing in pediatrics: Parents' considerations toward return of unsolicited findings for their child. *European Journal of Human Genetics, 24*(12), 1681–1687.

Djurdjinovic, L. (2011). Psychosocial counseling. In W. Uhlmann, J. Schuette, & B. Yashar (Eds.), *A guide to genetic counseling* (2nd ed). Hoboken, NJ: Wiley-Blackwell.

Dougherty, M. J., Lontok, K. S., Donigan, K., & McInerney, J. D. (2014). The critical challenge of educating the public about genetics. *Current Genetic Medicine Reports, 2*(2), 48–55.

Druker, H., Zelley, K., McGee, R. B., Scollon, S. R., Kohlmann, W. K., Schneider, K. A., & Schneider, K. W. (2017). Genetic counselor recommendations for cancer predisposition evaluation and surveillance in the pediatric oncology patient. *Clinical Cancer Research, 23*(13), e91–e97.

Duncan, R. E., & Young, M. A. (2013). Tricky teens: Are they really tricky or do genetic health professionals simply require more training in adolescent health? *Personalized Medicine, 10*(6), 589–600.

Ekstract, M., Holtzman, G. I., Kim, K. Y., Willis, S. M., & Zallen, D. T. (2017). Evaluation of a web-based decision aid for people considering the APOE genetic test for Alzheimer risk. *Genetics in Medicine*, *19*(6), 676.

Facio, F. M., Brooks, S., Loewenstein, J., Green, S., Biesecker, L. G., & Biesecker, B. B. (2011). Motivators for participation in a whole-genome sequencing study: Implications for translational genomics research. *European Journal of Human Genetics*, *19*(12), 1213−1217.

Haase, R., Michie, M., & Skinner, D. (2015). Flexible positions, managed hopes: The promissory bioeconomy of a whole genome sequencing cancer study. *Social Science & Medicine*, *130*, 146−153.

Haga, S. B., Mills, R., Pollak, K. I., Rehder, C., Buchanan, A. H., Lipkus, I. M., . . . Datto, M. (2014). Developing patient-friendly genetic and genomic test reports: Formats to promote patient engagement and understanding. *Genome Medicine*, *6*(7), 58.

Halverson, C. M., Clift, K. E., & McCormick, J. B. (2016). Was it worth it? Patients' perspectives on the perceived value of genomic-based individualized medicine. *Journal of Community Genetics*, *7*(2), 145−152.

Han, P. K., Umstead, K. L., Bernhardt, B. A., Green, R. C., Joffe, S., Koenig, B., . . . Biesecker, B. B. (2017). A taxonomy of medical uncertainties in clinical genome sequencing. *Genetics in Medicine*, *19* (8), 918−925.

Hayeems, R. Z., Babul-Hirji, R., Hoang, N., Weksberg, R., & Shuman, C. (2016). Parents' experience with pediatric microarray: Transferrable lessons in the era of genomic counseling. *Journal of Genetic Counseling*, *25*(2), 298−304.

Hooker, G. W., Ormond, K. E., Sweet, K., & Biesecker, B. B. (2014). Teaching genomic counseling: Preparing the genetic counseling workforce for the genomic era. *Journal of Genetic Counseling*, *23*(4), 445−451.

Hylind, R., Smith, M., Rasmussen-Torvik, L., & Aufox, S. (2017). Great expectations: patient perspectives and anticipated utility of non-diagnostic genomic-sequencing results. *Journal of Community Genetics*, 1−8.

Joseph, G., Pasick, R. J., Schillinger, D., Luce, J., Guerra, C., & Cheng, J. K. (2017). Information mismatch: Cancer risk counseling with diverse underserved patients. *Journal of Genetic Counseling*, 1−5.

Klitzman, R., Appelbaum, P. S., & Chung, W. (2013). Return of secondary genomic findings vs patient autonomy: Implications for medical care. *JAMA*, *310*(4), 369−370.

Kohler, J. N., Turbitt, E., & Biesecker, B. B. (2017). Personal utility in genomic testing: A systematic literature review. *European Journal of Human Genetics*, *25*(6), 662−668.

Krabbenborg, L., Schieving, J., Kleefstra, T., Vissers, L. E., Willemsen, M. A., Veltman, J. A., & van der Burg, S. (2016). Evaluating a counselling strategy for diagnostic WES in paediatric neurology: An exploration of parents' information and communication needs. *Clinical Genetics*, *89*(2), 244−250.

Levenseller, B. L., Soucier, D. J., Miller, V. A., Harris, D., Conway, L., & Bernhardt, B. A. (2014). Stakeholders' opinions on the implementation of pediatric whole exome sequencing: Implications for informed consent. *Journal of Genetic Counseling*, *23*(4), 552−565.

Linderman, M. D., Nielsen, D. E., & Green, R. C. (2016). Personal genome sequencing in ostensibly healthy individuals and the PeopleSeq Consortium. *Journal of Personalized Medicine*, *6*(2), 14.

Malek, J., Slashinski, M. J., Robinson, J. O., Gutierrez, A. M., Parsons, D. W., Plon, S. E., . . . McGuire, A. L. (2017). Parental perspectives on whole-exome sequencing in pediatric cancer: A typology of perceived utility. *JCO Precision Oncology*, *1*, 1−10.

Manolio, T. A., Chisholm, R. L., Ozenberger, B., Roden, D. M., Williams, M. S., Wilson, R., . . . Frazer, K. A. (2013). Implementing genomic medicine in the clinic: The future is here. *Genetics in Medicine*, *15*(4), 258−267.

Marron, J. M., DuBois, S. G., Bender, J. G., Kim, A., Crompton, B. D., Meyer, S. C., . . . Mack, J. W. (2016). Patient/parent perspectives on genomic tumor profiling of pediatric solid tumors: The Individualized Cancer Therapy (iCat) experience. *Pediatric Blood & Cancer*, *63*(11), 1974−1982.

McAllister, M., Wood, A. M., Dunn, G., Shiloh, S., & Todd, C. (2011). The Genetic Counseling Outcome Scale: A new patient-reported outcome measure for clinical genetics services. *Clinical Genetics*, *79*(5), 413−424.

Meiser, B., Irle, J., Lobb, E., & Barlow-Stewart, K. (2008). Assessment of the content and process of genetic counseling: A critical review of empirical studies. *Journal of Genetic Counseling, 17*(5), 434–451.

Merrill, S. L., & Guthrie, K. J. (2015). Is it time for genomic counseling? Retrofitting genetic counseling for the era of genomic medicine. *Current Genetic Medicine Reports, 3*(2), 57–64.

Miller, F. A., Hayeems, R. Z., Bytautas, J. P., Bedard, P. L., Ernst, S., Hirte, H., ... Winquist, E. (2014). Testing personalized medicine: Patient and physician expectations of next-generation genomic sequencing in late-stage cancer care. *European Journal of Human Genetics, 22*(3), 391–395.

Miller, V. A., Werner-Lin, A., Walser, S. A., Biswas, S., & Bernhardt, B. A. (2017). An observational study of children's involvement in informed consent for exome sequencing research. *Journal of Empirical Research on Human Research Ethics, 12*(1), 6–13.

Mollison, L., & Berg, J. S. (2017). Genetic screening: birthright or earned with age? *Expert Review of Molecular Diagnostics, 17*(8), 735–738.

Ormond, K. E. (2013). From genetic counseling to "genomic counseling". *Molecular Genetics & Genomic Medicine, 1*(4), 189–193.

Ormond, K. E., & Cho, M. K. (2014). Translating personalized medicine using new genetic technologies in clinical practice: The ethical issues. *Personalized Medicine, 11*(2), 211–222.

Ormond, K. E., Wheeler, M. T., Hudgins, L., Klein, T. E., Butte, A. J., Altman, R. B., ... Greely, H. T. (2010). Challenges in the clinical application of whole-genome sequencing. *The Lancet, 375* (9727), 1749–1751.

O'Rourke, P. P. (2013). Genomic medicine: Too great expectations? *Clinical Pharmacology & Therapeutics, 94*(2), 188–190.

Patenaude, A. F., & Schneider, K. A. (2017). Issues arising in psychological consultations to help parents talk to minor and young adult children about their cancer genetic test result: A guide to providers. *Journal of Genetic Counseling, 26*(2), 251–260.

Paul, J., Metcalfe, S., Stirling, L., Wilson, B., & Hodgson, J. (2015). Analyzing communication in genetic consultations—A systematic review. *Patient Education and Counseling, 98*(1), 15–33.

Pennington, J. W., Karavite, D. J., Krause, E. M., Miller, J., Bernhardt, B. A., & Grundmeier, R. W. (2017). Genomic decision support needs in pediatric primary care. *Journal of the American Medical Informatics Association, 24*(4), 851–856.

Phillips, K. A., Deverka, P. A., Sox, H. C., Khoury, M. J., Sandy, L. G., Ginsburg, G. S., ... Douglas, M. P. (2017). Making genomic medicine evidence-based and patient-centered: A structured review and landscape analysis of comparative effectiveness research. *Genetics in Medicine, 19*(10), 1081–1091.

Pichini, A., Shuman, C., Sappleton, K., Kaufman, M., Chitayat, D., & Babul-Hirji, R. (2016). Experience with genetic counseling: the adolescent perspective. *Journal of Genetic Counseling, 25*(3), 583–595.

Raymond, V. M., Gray, S. W., Roychowdhury, S., Joffe, S., Chinnaiyan, A. M., Parsons, D. W., & Plon, S. E. (2016). Germline findings in tumor-only sequencing: points to consider for clinicians and laboratories. *JNCI: Journal of the National Cancer Institute, 108*(4).

Resta, R., Biesecker, B. B., Bennett, R. L., Blum, S., Hahn, S. E., Strecker, M. N., & Williams, J. L. (2006). A new definition of genetic counseling: National Society of Genetic Counselors' task force report. *Journal of Genetic Counseling, 15*(2), 77–83.

Robinson, J. O., Carroll, T. M., Feuerman, L. Z., Perry, D. L., Hoffman-Andrews, L., Walsh, R. C., ... MedSeq Project Team. (2016). Participants and study decliners' perspectives about the risks of participating in a clinical trial of whole genome sequencing. *Journal of Empirical Research on Human Research Ethics, 11*(1), 21–30.

Roche, M. I., & Berg, J. S. (2015). Incidental findings with genomic testing: implications for genetic counseling practice. *Current Genetic Medicine Reports, 3*(4), 166–176.

Rolland, J., & Werner-Lin, A. (2006). *Families, health, and illness. Handbook of Health Social Work* (pp. 305–334). .

Rosell, A. M., Pena, L. D., Schoch, K., Spillmann, R., Sullivan, J., Hooper, S. R., ... Shashi, V. (2016). Not the end of the odyssey: Parental perceptions of whole exome sequencing (WES) in pediatric undiagnosed disorders. *Journal of Genetic Counseling, 25*(5), 1019–1031.

Ross, L. F., Saal, H. M., David, K. L., Anderson, R. R., & American Academy of Pediatrics. (2013). Technical report: Ethical and policy issues in genetic testing and screening of children. *Genetics in Medicine, 15*(3), 234—245.

Roter, D., Ellington, L., Erby, L. H., Larson, S., & Dudley, W. (2006). *The genetic counseling video project (GCVP): Models of practice,* . American Journal of Medical Genetics Part C: Seminars in Medical Genetics (Vol. 142, pp. 209—220). Wiley Subscription Services, Inc., A Wiley Company, No. 4.

Roter, D. L., Erby, L. H., Larson, S., & Ellington, L. (2007). Assessing oral literacy demand in genetic counseling dialogue: Preliminary test of a conceptual framework. *Social Science & Medicine, 65*(7), 1442—1457.

Rubel, M. A., Werner-Lin, A., Barg, F. K., & Bernhardt, B. A. (2017). Expert knowledge influences decision-making for couples receiving positive prenatal chromosomal microarray testing results. *Culture, Medicine, and Psychiatry, 41*(3), 382—406.

Samuel, G. N., Dheensa, S., Farsides, B., Fenwick, A., & Lucassen, A. (2017). Healthcare professionals' and patients' perspectives on consent to clinical genetic testing: Moving towards a more relational approach. *BMC Medical Ethics, 18*(1), 47.

Sanderson, S. C., Linderman, M. D., Suckiel, S. A., Diaz, G. A., Zinberg, R. E., Ferryman, K., ... Schadt, E. E. (2016). Motivations, concerns and preferences of personal genome sequencing research participants: Baseline findings from the HealthSeq project. *European Journal of Human Genetics, 24*(1), 14—20.

Schmidlen, T., Sturm, A. C., Hovick, S., Scheinfeldt, L., Roberts, J. S., Moor, L., & Sweet, K. (2018). Operationalizing the reciprocal engagement model of genetic counseling practice: A framework for the scalable delivery of genomic counseling and testing. *Journal of Genetic Counseling, 27*(5), 1111—1129.

Schnitzler, L., Smith, S. K., Shepherd, H. L., Shaw, J., Dong, S., Carpenter, D. M., ... Dhillon, H. M. (2017). Communication during radiation therapy education sessions: The role of medical jargon and emotional support in clarifying patient confusion. *Patient Education and Counseling, 100*(1), 112—120.

Scollon, S., Bergstrom, K., Kerstein, R. A., Wang, T., Hilsenbeck, S. G., Ramamurthy, U., ... McCullough, L. B. (2014). Obtaining informed consent for clinical tumor and germline exome sequencing of newly diagnosed childhood cancer patients. *Genome Medicine, 6*(9), 69.

Skinner, D., Raspberry, K. A., & King, M. (2016). The nuanced negative: Meanings of a negative diagnostic result in clinical exome sequencing. *Sociology of Health & Illness, 38*(8), 1303—1317.

Skinner, D., Roche, M. I., Weck, K. E., Raspberry, K. A., Foreman, A. K., Strande, N. T., ... Henderson, G. E. (2018). "Possibly positive or certainly uncertain": Participants' responses to uncertain diagnostic results from exome sequencing. *Genetics in Medicine, 20*, 313—319.

Tabor, H. K., Jamal, S. M., Yu, J. H., Crouch, J. M., Shankar, A. G., Dent, K. M., ... Bamshad, M. J. (2017). My46: A Web-based tool for self-guided management of genomic test results in research and clinical settings. *Genetics in Medicine, 19*(4), 467.

Vassy, J. L., McLaughlin, H. L., MacRae, C. A., Seidman, C. E., Lautenbach, D., Krier, J. B., ... Rehm, H. L. (2015). A one-page summary report of genome sequencing for the healthy adult. *Public Health Genomics, 18*(2), 123—129.

Veach, P. M., Bartels, D. M., & LeRoy, B. S. (2007). Coming full circle: a reciprocal-engagement model of genetic counseling practice. *Journal of Genetic Counseling, 16*(6), 713—728.

Wakefield, C. E., Hanlon, L. V., Tucker, K. M., Patenaude, A. F., Signorelli, C., McLoone, J. K., & Cohn, R. J. (2016). The psychological impact of genetic information on children: A systematic review. *Genetics in Medicine, 18*(8), 755—762.

Walser, S. A., Werner-Lin, A., Mueller, R., Miller, V. A., Biswas, S., & Bernhardt, B. A. (2017). How do providers discuss the results of pediatric exome sequencing with families? *Personalized Medicine, 14*(5), 409—422.

Walser, S. A., Werner-Lin, A., Russell, A., Wapner, R. J., & Bernhardt, B. A. (2016). "Something extra on chromosome 5": Parents' Understanding of positive prenatal chromosomal microarray analysis (CMA) results. *Journal of Genetic Counseling, 25*(5), 1116—1126.

Watson, S. L., Hayes, S. A., & Radford-Paz, E. (2011). Diagnose me please!": A review of research about the journey and initial impact of parents seeking a diagnosis of developmental disability for their child. *International Review of Research in Developmental Disabilities, 41*, 31—72.

Werner-Lin, A., McCoyd, J. L., & Bernhardt, B. A. (2016). Balancing genetics (science) and counseling (art) in prenatal chromosomal microarray testing. *Journal of Genetic Counseling, 25*(5), 855−867.

Werner-Lin, A., Merrill, S. L., & Brandt, A. C. (2018). Talking with children about adult-onset hereditary cancer risk: A developmental approach for parents. *Journal of Genetic Counseling, 27*(3), 533−548.

Werner-Lin, A., Tomlinson, A., Miller, V., & Bernhardt, B. A. (2016). Adolescent engagement during assent for exome sequencing. *AJOB Empirical Bioethics, 7*(4), 275−284.

Werner-Lin, A., Zaspel, L., Carlson, M., Mueller, R., Walser, S. A., Desai, R., & Bernhardt, B. A. (2018). Gratitude, protective buffering, and cognitive dissonance: How families respond to pediatric whole exome sequencing in the absence of actionable results. *American Journal of Medical Genetics Part A, 176*(3), 578−588.

Wicklund, C., & Trepanier, A. (2014). Adapting genetic counseling training to the genomic era: more an evolution than a revolution. *Journal of Genetic Counseling, 23*(4), 452−454.

Wynn, J., Lewis, K., Amendola, L., Bernhardt, B. A., Biswas, S., Joshi, M., & Scollon, S. (2018). Clinical providers' experiences with returning results from genomic sequencing. *BMC Medical Genomics, 11* (1), 45.

Yu, J.-H., Appelbaum, P. S., Brothers, K. B., Joffe, S., Kauffman, T. L., Koenig, B. A., . . . Wilfond, B. S. Consent for clinical genome sequencing: Considerations from the Clinical Sequencing Exploratory Research Consortium (under review).

CHAPTER 9

Genome Sequencing in Pediatrics: Ethical Issues

Candice Cornelis[1,2,]* and Roel H.P. Wouters[3,]*
[1]Department of Genetics, University Medical Center Utrecht, Utrecht, The Netherlands
[2]Ethics Institute, Utrecht University, Utrecht, The Netherlands
[3]Department of Medical Humanities, Julius Center, University Medical Center Utrecht, Utrecht, The Netherlands

INTRODUCTION

In genomics and genetics, psychological, medical, and ethical issues are substantially intertwined and have therefore rightly been discussed in conjunction with one another. Amongst other issues, protecting the patient's autonomy has received a great deal of attention, both in the relevant bioethics literature and in clinical practice, in developing return of results policies. The fundamental idea is that information about one's genetic constitution is inherently personal and should therefore only be disclosed with the individual's consent. Consequently, a great part of the ethics literature in the field of genetics and genomics has focused on how the informed consent process ought to be structured, including the way in which choices should be offered to patients (Appelbaum et al., 2014; Wolf et al., 2008). This task has become increasingly complex as a result of the introduction of comprehensive testing modalities such as whole genome sequencing (WGS) or whole exome sequencing (WES). These genomic tests have the ability to produce not just information that is intentionally sought after but also information that is generated as a byproduct of using a nonspecific test: unsolicited findings (UFs). Alternatively, UFs are sometimes referred to as incidental or secondary findings (Tan et al., 2017). Although each term has previously been defined as conveying a slightly different message, these terms are used interchangeably in the literature as well as in practice (for consistency, UFs will be used throughout this chapter). Ensuring that patients are able to make an informed choice is complicated due to the vast number and heterogeneity of UFs that could potentially be discovered (Bredenoord, Kroes, Cuppen, Parker, & van Delden, 2011; Christenhusz, Devriendt, & Dierickx, 2013).

In children, this process runs into additional difficulties because children are not yet capable of making autonomous decisions (Borry, Stultiëns, Nys, Cassiman, &

* Both authors contributed equally.

Clinical Genome Sequencing
DOI: https://doi.org/10.1016/B978-0-12-813335-4.00009-X

Dierickx, 2006). Similar difficulties arise throughout pediatric medicine as a whole and generally these issues are addressed by allowing parents to decide for or on behalf of their children. Most of the time, this strategy is rather uncontroversial as deferring medical treatment until a child is able to decide for itself is usually not an option, particularly if treatment needs to commence immediately, during childhood, in order to be effective. Ethical dilemmas are usually confined to situations in which the parents' decision runs contrary to the (alleged) interests of the child. Nonetheless, genomic testing can also yield information that only becomes relevant after the child has reached the age of majority: mutations that predispose the individual to adult-onset disease or carrier status (Botkin et al., 2015; Clayton et al., 2014; Cornelis et al., 2016). Hence, in the case of these types of findings, deferring disclosure of UFs is a reasonable possibility. In this chapter, several ethical aspects of UFs from pediatric genomic testing will be highlighted. After briefly mapping international policies and recommendations, the relevant evidence from empirical studies will be summarized. Consequently, in the discussion section, this empirical evidence will be put into perspective and linked to normative theories. The discussion will also provide an analysis of the moral values that play a pivotal role in formulating guidelines for managing UFs and put forward a concrete scheme of what such guidelines may look like.

CURRENT POLICY GUIDELINES FOR GENOMIC TESTING IN MINORS

In the international policy debate, various proposals have been put forth regarding return of results from next-generation sequencing techniques. Many policies suggest distinguishing between clinically relevant UFs based on whether the finding carries health-related consequences for those involved. In addition, these policies distinguish between medically actionable or nonactionable UFs, based on whether the likelihood of aforementioned consequences can be influenced by therapy or prevention (Wolf et al., 2008). Finally, UFs are divided into childhood-onset and adult-onset, depending at what point a condition reveals itself. Some findings may have reproductive significance, which are usually described as a separate category (although strictly speaking reproductive decisions can be medical actions too) (Bredenoord, Onland-Moret, & Van Delden, 2011). Different guidelines offer different answers as to what types of findings may or may not be disclosed and under what conditions, that is, should parents (or other legal representatives of the child) be offered choices or not. The answers differ in how conflicts should be addressed between the various interests of children, parents, and the family at large.

While there appears to be broad consensus on UFs that offer life-saving medical actionability in childhood being returned in the case of children incapable of participating in any future decision-making process (e.g., intellectual disability), divergence exists on how to handle findings with adult onset—whether or not medically actionable. Where some policies advise against disclosing such findings, others offer more parental discretion.

Likewise, policies differ regarding the role of adolescents who have not yet reached full maturity but are nonetheless capable of expressing their own views. For example, Holm et al.'s preference-based model (Holm et al., 2014), developed for return of genomic sequencing results in the research context, acknowledges that adolescents are capable of participating in the decision-making alongside parents, and allows parents and adolescents to individually indicate which UFs they would or would not like to receive prior to sequencing. If a UF is discovered, an oversight board takes parental and adolescent preferences into account in their deliberations whether to disclose/withhold the finding(s), along with moral considerations. This approach recognizes that the family dynamic may warrant different decisions regarding return of individual results, insofar as parents and/or adolescents may have different preferences for findings within the same category and there are potential health/reproductive consequences for other siblings and/or relatives. The model aims to respect parents' and adolescents' preferences optimally. At the same time, the model acknowledges that although participants should be able to indicate preferences, it remains a question whether these should be respected. This is because the model also aims to prevent harm, including harm related to interference with a child's future autonomy. Furthermore, the model allows children to receive results pertaining to themselves that were initially withheld once they reach the age of majority (when they are still enrolled in the research).

Additionally, there is the question of whether UFs should be filtered out as much as possible, assuring maximum focus on the clinical indication for sequencing, or whether a subset of actionable mutations ought to be routinely screened for (Wouters et al., 2018). Policy recommendations published in 2013 by the European Society of Human Genetics advocate a filtered approach, directing labs and clinicians to first use a targeted approach to sequencing to limit the chances of revealing UFs as much as possible (Van El, Cornel, et al., 2013; Van El, Dondorp, De Wert, & Cornel, 2013). In practice, this amounts to the use of gene panels compiled for a specific medical indication, for instance intellectual disability or cancer, and then (if necessary) gradually widening the search by analyzing sequencing data on other genes. In contrast, the American College of Medical Genetics and Genomics (ACMG) advocates for routine screening for actionable UFs. In these policy recommendations, labs are directed to conduct additional analysis of certain listed genes (a list which has been revised), regardless of the specific medical indication for sequencing (Green et al., 2013; Kalia et al., 2017). The list includes pathogenic variants for conditions with childhood onset, and for adult-onset conditions that lack medical actionability in childhood. Initially parents/competent patients were required to consent to this search before sequencing could commence. Since 2014, however, persons undergoing sequencing may opt-out of the additional search of all listed genes. This constitutes an all-or-nothing approach:

parents/patients must either forego gaining knowledge of any UF in a listed gene or accept any UF found in a listed gene. It is not possible to opt-out of receiving certain findings versus others (ACMG Board of Directors, 2015). Although the ACMG refers to this additional search as a search for "incidental findings," it is worth noting that in a strict sense these findings are not incidental, since they are ones that are being intentionally sought after or "hunted" for. For this reason, some prefer to refer to these types of results as secondary findings (Jarvik et al., 2014).

EMPIRICAL ETHICS RESEARCH ON GENOMIC TESTING IN MINORS

Several empirical studies have previously explored preferences and attitudes of parents regarding UFs discovered in their children's genomes. Initially, the scholarship on this subject consisted predominantly of hypothetical scenarios being proposed to parents in order to elicit their responses. In the last couple of years, however, a growing number of studies have been published that focus on parents with children having actual sequencing experience (Anderson et al., 2017; Clift et al., 2015; Cornelis et al., 2016; Fernandez et al., 2014; Rigter et al., 2014; Sapp et al., 2014; Shahmirzadi et al., 2014). From these studies, a broad idea about how parents view UFs can be constructed. In general, parents are keen to receive feedback on a wide range of potential UFs pertaining to their children's health (Cornelis et al., 2016; Sapp et al., 2014). This conclusion is congruous with observations from studies that surveyed or interviewed the general public (Mackley, Fletcher, Parker, Watkins, & Ormondroyd, 2017; Middleton et al., 2016). A variety of reasons for wanting to receive these findings were identified. What stands out is the desire to protect one's child from being harmed in the future by hereditary diseases. Parents are particularly ardent about the type of results commonly designated as actionable findings, mutations predisposing children to diseases that could be treatable or preventable. However, many parents' preferences are by no means limited to these actionable mutations, since multiple studies have demonstrated a widely shared parental interest in mutations that predispose for nontreatable conditions (Fernandez et al. 2014; Kleiderman et al., 2014; Levenseller et al., 2014; Sapp et al., 2014). This is also true for carrier status of recessive disorders, information that will not have any consequences for the health of the child that is tested but may influence his or her reproductive decisions later in life. Although carrying a mutation that causes recessive disorders becomes directly relevant to the child only after reaching maturity, parents may have several reasons for wanting to receive information on nonactionable conditions, actionable adult-onset conditions, and recessive disorders. These reasons range from being alerted that they could themselves be at risk for hereditary diseases to having the possibility to discuss risks for future diseases with their children. Also, information about mutations that cause recessive diseases could affect the parent's decision to have more children or could encourage them to explore reproductive technologies such as preconception screening and preimplantation genetic diagnosis (Mackley et al., 2017). In addition to wanting to learn about UFs for medical reasons, some parents value genetic or genomic findings simply for the sake of being informed, for example because they feel comfortable having as much information on their child's health as possible and because they want to be prepared for the occurrence of certain diseases in the future (Cornelis et al., 2016; Levenseller et al., 2014; Sapp et al., 2014). Although some studies have revealed that parents see the child's future autonomy as a consideration favoring declining such results (Cornelis et al., 2016), many parents seem to regard themselves as legitimate recipients of information pertaining to their child's health. In general, parents appear to claim the right to decide which of the UFs they want to convey to their children. The observation that most parents see themselves as the primary gatekeepers of their children's genetic information stands in contrast with the emphasis that many guidelines put on the role of the professional (Kleiderman et al., 2014).

It should be noted, however, that the contexts in which these empirical data are assembled are heterogeneous. WES or WGS may be used either for diagnostic purposes or for therapeutic purposes. Diagnostic sequencing takes place predominantly within the clinical genetics department, while sequencing aimed at identifying biomarkers is also done within other fields of pediatrics. For example, pediatric oncologists increasingly use genome sequencing to find potential molecular targets for therapy (McCullough et al., 2016). Patients and their parents who navigate through sequencing decisions in these different contexts have distinct backgrounds, disease experiences, clinical pathways, and treating physicians, all of which may well affect their views on UFs. In addition to this context heterogeneity, studies available in the literature are also qualitatively different regarding the characteristics of the children that were being studied. In particular, the participants include parents of children with variable age and cognitive developmental status. These factors are of paramount importance to policies governing disclosure of UFs because it matters whether or not the child is able to express its own preferences or will be able to do so in the future (Cornelis et al., 2016). For example, deferring feedback on mutations that predispose for adult-onset conditions until the child reaches maturity in order to protect the child's autonomous decision-making in the future may be pointless for some children who undergo sequencing as a diagnostic tool for mental retardation.

Another gap in the current literature on UFs in pediatrics is constituted by children's own preferences and experiences. Most published studies explore parents' views, opinions, worries, etc. about return of UFs. Studies aimed at elucidating the child's own views on return of UFs are scarce (Hufnagel, Martin, Cassedy, Hopkin, & Antommaria, 2016). In part, this is only natural because many of these children are too young to be interviewed. However, young children whose DNA has been sequenced could be included in longitudinal studies that allow researchers to study how patients look back on their experiences and how they appraise choices made by their parents during the sequencing procedure. Longitudinal studies could also shed light on long-term consequences of receiving an UF. Several positive and negative consequences of receiving genomic information have been mentioned in the literature. These consequences could be positive, for example, by expediting future diagnoses, introducing the possibility to prevent diseases or the ability to make better-informed life-planning decisions. However, there could also be negative consequences, including a psychological burden of knowing to be at risk for diseases and social burdens arising from the way society treats individuals known to carry a specific type of mutation (e.g., limited insurability). Currently, the literature on both positive and negative consequences remains largely speculative. Longitudinal studies are needed to find out whether and to what extent these effects actually occur. Until these results become available, professionals have to rely on data from studies that focus on the impact of either disclosing genomic UFs to adults or communicating results from traditional genetic tests that analyze specific genes. For example, studies on the effects of genetic testing for *BRCA* mutations (hereditary breast and ovarian cancer) do not support the worry that conveying cancer susceptibility causes lasting psychological distress or anxiety (Hirschberg et al., 2015).

TRANSLATING EMPIRICAL RESULTS INTO NORMATIVE POLICIES

Empirical research is increasingly being used in contemporary bioethics to inform normative debates with views of those who have first-hand experience regarding the issue at stake. Yet, even if new studies fill all the gaps in the empirical body of knowledge that have been summarized above, that would not provide ready-made solutions for many of the ethical issues that are discussed here. There is a gap between the empirical and the normative that cannot be bridged by more empirical research. Studies aimed at elucidating patient preferences towards UFs of genomic testing demonstrate that individuals typically hold quite diverse views on this matter (Mackley et al., 2017). This heterogeneity raises the question as to which preferences should be accommodated by healthcare policies. In addition, specifically in child cases, adolescents and their parents may have opposing preferences, which reflects that they may have potentially conflicting interests that are at stake. Professionals have to navigate through protecting these different interests. Potentially conflicting interests that professionals should normally protect but which they cannot always simultaneously protect when dealing with UFs include the child's (future) autonomy, the child's health status, the parents' health status, the parents' autonomy, and parental discretion to decide on behalf of their (underage) children. Protecting the child's current health status is often-times a prerequisite for ensuring future autonomy—should the child be able to become autonomous—if (preventive) treatment decisions cannot be safely postponed until a later point in time. A prominent type of conflict in genetics and genomics is preserving the child's future autonomy versus protecting the parents' autonomy and the parental discretion to decide on behalf of their (underage) children. An important part of ensuring the child's future autonomy includes refraining from making certain decisions about crucial facets of children's lives that could be deferred until adulthood, including decisions about genetic risk. Policies to defer these decisions anticipate that children, once they have become adults, may prefer not to know whether they carry a pathogenic mutation or not. In the remainder of this chapter, we will focus on this aspect of future autonomy in the debates on return of UFs in pediatric cases.

THE ETHICAL DEBATE RECONSIDERED

The ethical debate on pediatric UFs has centered around the issue of how to handle pathogenic variants that predispose children to diseases that reveal themselves in adulthood (adult-onset conditions) and on carrier status for recessive conditions. Where some have argued that information on these types of findings (usually) does not have clinical consequences in childhood and feedback can therefore be postponed until the child reaches the age of competence, others have taken a more permissive approach, allowing more discretion to parents over receiving these findings at present.

Those in favor of postponing feedback on adult-onset conditions usually invoke Feinberg's "right to an open future" argument (Bredenoord, De Vries, & Van Delden, 2013; Bredenoord, De Vries, & van Delden, 2014; Davis, 1997). According to Feinberg, a right to an open future is a blanket term for a collection of autonomy rights that children cannot exercise now, but will be able to once competent. However, these rights can be violated at present. This is why a right to an open future is seen as a "right in trust": persons that are responsible for taking care of children (parents, physicians, etc.) must act as trustees on children's behalf in such a way that those children can lead a life of their own choosing as autonomous adults (Feinberg, 1980).

Hence, professionals should keep possibilities concerning testing for diseases open to children so that they can make their own decisions later in life. This argument was developed in a debate on religious freedom in relation to compulsory education, but later became a common argument in genetics. However, how the content of this right ought to be understood, that is, what it aims to protect, and how it can be ethically justified, is still a subject of debate (Wouters, Cornelis, Newson, Bunnik, & Bredenoord, 2017).

Two interpretations of Feinberg's right to an open future have been distinguished in the literature. On what has been referred to as the "maximization interpretation," Feinberg's right is taken to mean that children, once competent, should have as many options open to them as possible to choose amongst. Yet many if not most authors do not accept this maximization interpretation for a variety of reasons, including that it places unreasonable demands on parents as well as the state (Archard, 2016). In addition, it views the aim of parenting in terms of option maximization for children, which means parents would have to sacrifice many other goals they may have in their lives as being unworthy of pursuit in order to ensure option maximization. Some have also argued that the idea of a maximally open future is incoherent, since in choosing one option, we automatically close off other options. Instead, most authors uphold a more modest interpretation that encompasses both positive and negative moral duties toward children: children should be aided in their development of certain capacities and skills and only certain key options should be kept open to them that are *reasonable* to keep open (Archard, 2016).

Despite these discussions about the justification and content of a right to an open future, many authors working in genetics have advocated an interpretation of the concept, which resulted in a strong, but not complete consensus that children should not be tested for adult-onset conditions until they are able to decide whether or not to receive this information.

In opposition to Feinberg's standpoint (on children's rights in general), Ruddick developed what has sometimes been referred to as a family-based account of what parents owe to their children that he called the "Life Prospects Principle" (Ruddick, 1979). According to this principle parents should be seen as providers of "Life Prospects" rather than, as the Feinbergian account contends, trustees of children's rights that children themselves cannot exercise. Where trustees must simply act in accordance with what has been bestowed upon them to entrust, that is, future possibilities, without taking their own or other family members' interests into account, providers should focus on actively searching for possible ways of leading one's life (life possibilities) that a child, once autonomous, could accept. On Ruddick's view, parents are allowed to let their own preferences be guiding to some extent in deciding what life possibilities they want to provide their child with.

One of the main reasons Ruddick argues for parents as providers for their child is because he conceives of children as being as a kind of product of parents, albeit a distinct class of products, that will gradually develop into autonomous beings, given the right kind of nurturing. As a child develops into an autonomous being, his or her interests and decision-making abilities become increasingly separated from parents and other family members, even if it is still complicated to pinpoint where childhood stops and autonomous adulthood starts and thus when parenthood, in the sense of acting as a provider of life possibilities, tapers off in importance. Ruddick therefore argues against views that stipulate that it is possible to neatly differentiate between children's interests and parents' interests—since these will be, at least to some extent, intertwined with one another. His account affords more space to parents to make their own decisions about what life prospect to provide their child with, since the criteria for deeming a life prospects morally good are that the child could accept that possibility and make it their own in the future without relying on continuous help from their parents. As Bredenoord et al. (2014) have pointed out, Ruddick's Life Prospects account could justify disclosing a broader range of genetic mutations, including ones for carrier status and certain adult-onset conditions lacking (preventive) medical action.

However, whether or not Ruddick's account allows parents to gain knowledge of these types of findings is debatable. According to another interpretation of Ruddick:

> *The major thrust of Ruddick's principle, then, is to resist ideas that parents are their children's masters or their servants, their guardians or their trustees, and to replace these ideas with the view that the most distinctive work of parenthood is life-giving. This includes providing the child with the possibility of leading a life of his or her own, a life that, ideally at least, is mutually acceptable to the child and parents alike.*

> *... ... as Ruddick argues, parents have an obligation to provide the child with a number of choices as to various ways of life, including that upheld by the family itself... (Phillips, 1986, p. 179).*

According to Phillips, parents have an obligation to introduce their child to different life prospects that may indeed rule out one another if one or the other is realized. This includes introducing them to certain possibilities that the parents themselves might not prefer. Children need to have these choices in order to optimally develop their reflective capacities. Developing such capacities is a necessary component of becoming an autonomous adult who can go on to seek out their own purposes in life. Phillips does not argue that according to Ruddick's view, parents' preferences are subordinate to their child's interests, at least not by definition, but he assumes that most parents would prefer their child to develop reflective capacities at least to a certain extent.

Applied to genomics, Phillips's account of the Feinberg—Ruddick debate could be taken to imply that returning certain information about adult-onset conditions or carrier-status should be postponed until the child can decide for themselves about whether or not to receive this information. For, if such findings are returned, then the child has no way of knowing what a childhood or even adulthood life would look like without this information. Postponing the choice still allows the child to choose whether or not to realize the prospect of knowing versus not knowing.

On this latter interpretation of Ruddick's Life Prospect Principle, Ruddick is not a radical conservative who proposes full parental discretion but rather a moderate who states that parents should present the child with more than one future prospect that their children may pursue once grown-up. Even though parents are rightfully seen as shapers of parts of their child's future, they must also provide life prospects that are also reasonable from the child's own future perspective.

When applied to genomics, the nuanced interpretation of Ruddick yield outcomes that are strikingly similar to the moderate interpretation of Feinberg, that is, the interpretation that future autonomous choices should not be maximized but be preserved to a reasonable extent. This observation introduces a novel insight to the debate on UFs in pediatrics, as it clarifies that also from a Ruddickian perspective, leaving different options or prospects open is a valuable or even preferable strategy. Feinberg's as well as Ruddick's philosophical ideas on childhood and parenthood seem to imply that parents should postpone decision-making regarding certain types of UFs so that the child can decide for himself/herself—such postponement secures the child's future autonomy.

Before turning to a sketch of how this moderate position could be translated into concrete guidelines, several additional factors need to be taken into account. First, the sequencing workflow in a pediatric setting frequently uses a specific modality called trio analysis as a diagnostic tool to evaluate congenital abnormalities. In a trio analysis, the parents' genome is sequenced alongside the child's genome to detect de novo mutations that may have caused abnormalities in the child. This method is useful in the case of de novo mutations whose relation to the abnormality was previously unknown and as a means to filter out many candidates that might have led to the disease but are unlikely to be the cause of the disease because an unaffected parent carries the same variant. The widespread use of trio analyses forms a qualitatively and morally different situation from a sequencing procedure in which only the child's DNA is being sequenced. In the latter setting, a professional that discovers a particular mutation can only infer that one of the parents is likely to carry that mutation. By contrast, in the case of a trio analysis the parents' genomic data are on the same computer, ready to be checked in the blink of an eye. On theoretical grounds, having the genomic data of the parents available could be a sound possibility to circumvent the quandary between protecting the child's future autonomy and the parents' health. For example, one could simply tell the parents that they carry an actionable mutation without informing them of the carrier status of their child. In practice, however, this solution can be more difficult to achieve as a result of the method by which genomic data generated by trio analyses are analyzed. One of these methods is to initially identify aberrations in the child's DNA and, consequently, to check whether this aberration is also present in the DNA of the parents. The remaining part of the parental DNA is filtered out and remains invisible. Hence, in such a procedure, an UF cannot be returned to the parents without informing them that the mutation is also carried by the child because in the case the mutation was not detected in the child's DNA the mutation would not have been detected at all.

A second relevant consideration in deciding whether or not to disclose certain UFs at present or save them for future disclosure is what the means of information storage are. At present, storage of data for future use is not always possible in a clinical context—either due to a lack of a so-called "shadow storage system" or because of legal aspects of what is required to be noted in medical records. If UFs cannot be stored for later disclosure, then there may be a stronger case for disclosing certain information immediately following sequencing, especially if the information in question constitutes medically actionable information.

Third, juxtaposing the child's autonomy and the parents' health as two mutually opposing interests may be a theoretical abstraction of values that are deeply intertwined in practice. In reality, parents have a vital role in fostering and nurturing the child's future autonomy. For parents to succeed in this task, good health is paramount. Nondisclosure of actionable findings may therefore be counterproductive because living with an ill parent (or even losing one) may have detrimental effects on the child's autonomy. Moreover, these harmful effects could greatly exceed the positive effects of nondisclosure.

From this analysis, some tentative conclusions can be formulated, even though the debate on how to handle UFs in children is still ongoing. In the case of a trio analysis, disclosing UFs to the child may become inevitable upon detecting a pathogenic variant in one of the parents' DNA as a result of the way in which information is analyzed. Although conveying results that are actionable and adult-onset (especially in the case of a pathogenic variant in a gene that rarely occurs de novo) appears to violate the child's (future) autonomy in that they can no longer decide whether or not to receive this information, we think that preserving the good health of the parents is actually supportive of the child being able to develop into an autonomous being and not simply warranted by virtue of upholding a moral duty to protect the parents' health. Finally, the policy toward carrier status of autosomal recessive mutations and mutations predisposing for nonactionable conditions that are adult-onset in the child's DNA depend on the specific situation and these decisions should be left to a committee. Based on the insight that postponing these decisions is vindicated by reasonable interpretations of both the right to an open future argument and the Life Prospects Principle, the ethically preferred way to go about such findings would be to postpone the decision in order to leave the option of rejecting this type of information open to the child. However, some situations might warrant disclosure, such as when it is possible that the information may provide parents with pivotal reproductive options.

CONCLUSION

Various guidelines are available for how UFs should be managed by healthcare professionals, but these contain diverging recommendations on how these findings should be managed. A growing number of empirical studies elucidating parental preferences toward UFs have been published in the literature, which demonstrate that many parents are interested in learning about UFs discovered in their children's DNA. This poses a pivotal moral concern of how to navigate between upholding the child's future autonomy versus the parents' autonomy to decide on behalf of their children. These values touch upon underlying views on the rights of the child and the role of parents in their upbringing. As mentioned above, multiple interpretations of both Feinberg and Ruddick are available. However, as we have argued, it is also possible to understand both positions as attempts to justify a middle ground in which certain options should be kept open to a reasonable extent and we sketched various avenues of how this general idea could be translated into proposals for dealing with pediatric UFs in daily practice. This includes the suggestion that it is warranted to return UFs for adult-onset, medically actionable conditions in a trio-analysis situation and that UFs for carrier status for autosomal recessive conditions and medically nonactionable UFs generally should not be eligible for disclosure, save circumstances where this knowledge may provide parents with important reproductive options.

REFERENCES

ACMG Board of Directors. (2015). ACMG policy statement: Updated recommendations regarding analysis and reporting of secondary findings in clinical genome-scale sequencing. *Genetics in Medicine*, *17*, 68−69.

Anderson, J. A., Meyn, M. S., Shuman, C., Shaul, R. Z., Mantella, L. E., Szego, M. J., ... Hayeems, R. Z. (2017). Parents perspectives on whole genome sequencing for their children: Qualified enthusiasm? *Journal of Medical Ethics*, *43*, 535−539.

Appelbaum, P. S., Waldman, C. R., Fyer, A., Klitzman, R., Parens, E., Martinez, J., ... Chung, W. K. (2014). Informed consent for return of incidental findings in genomic research. *Genetics in Medicine*, *16*, 367−373.

Archard, D. W. (2016). Children's rights. *The Stanford Encyclopedia of Philosophy*. Available from https://plato.stanford.edu/archives/sum2016/entries/rights-children/.

Borry, P., Stultiëns, L., Nys, H., Cassiman, J. J., & Dierickx, K. (2006). Presymptomatic and predictive genetic testing in minors: A systematic review of guidelines and position papers. *Clinical Genetics*, *70*, 374−381.

Botkin, J. R., Belmont, J. W., Berg, J. S., Berkman, B. E., Bombard, Y., Holm, I. A., ... McInerney, J. D. (2015). Points to consider: Ethical, legal, and psychosocial implications of genetic testing in children and adolescents. *The American Journal of Human Genetics*, *97*, 6−21.

Bredenoord, A. L., De Vries, M. C., & Van Delden, J. J. M. (2013). Next-generation sequencing: Does the next generation still have a right to an open future? *Nature Reviews Genetics*, *14*, 306.

Bredenoord, A. L., De Vries, M. C., & van Delden, H. (2014). The right to an open future concerning genetic information. *The American Journal of Bioethics*, *14*, 21−23.

Bredenoord, A. L., Kroes, H. Y., Cuppen, E., Parker, M., & van Delden, J. J. (2011). Disclosure of individual genetic data to research participants: The debate reconsidered. *Trends in Genetics*, *27*, 41−47.

Bredenoord, A. L., Onland-Moret, N. C., & Van Delden, J. J. M. (2011). Feedback of individual genetic results to research participants: In favor of a qualified disclosure policy. *Human Mutation*, *32*, 861−867.

Christenhusz, G. M., Devriendt, K., & Dierickx, K. (2013). To tell or not to tell? A systematic review of ethical reflections on incidental findings arising in genetics contexts. *European Journal of Human Genetics*, *21*, 248−255.

Clayton, E. W., McCullough, L. B., Biesecker, L. G., Joffe, S., Ross, L. F., Wolf, S. M., & for the Clinical Sequencing Exploratory Research (CSER) Consortium Pediatrics Working Group. (2014). Addressing the ethical challenges in genetic testing and sequencing of children. *The American Journal of Bioethics*, *14*, 3−9.

Clift, K. E., Halverson, C. M., Fiksdal, A. S., Kumbamu, A., Sharp, R. R., & McCormick, J. B. (2015). Patients' views on incidental findings from clinical exome sequencing. *Applied & Translational Genomics, 4,* 38−43.

Cornelis, C., Tibben, A., Dondorp, W., Van Haelst, M., Bredenoord, A. L., Knoers, N., ... van Summeren, M. (2016). Whole-exome sequencing in pediatrics: Parents' considerations toward return of unsolicited findings for their child. *European Journal of Human Genetics, 24,* 1681−1687.

Davis, D. S. (1997). Genetic dilemmas and the child's right to an open future. *Hastings Center Report, 27,* 7−15.

Feinberg, J. (1980). A child's right to an open future. In W. Aiken, & H. LaFollette (Eds.), *Whose child? Parental rights, parental authority and state power* (pp. 124−153). Totowa, NJ: Littlefield, Adams, and Co.

Fernandez, C. V., Bouffet, E., Malkin, D., Jabado, N., O'Connell, C., Avard, D., ... McMaster, C. R. (2014). Attitudes of parents toward the return of targeted and incidental genomic research findings in children. *Genetics in Medicine, 16,* 633−640.

Green, R. C., Berg, J. S., Grody, W. W., Kalia, S. S., Korf, B. R., Martin, C. L., ... Biesecker, L. G. (2013). ACMG recommendations for reporting of incidental findings in clinical exome and genome sequencing. *Genetics in Medicine, 15,* 565−574.

Hirschberg, A. M., et al. (2015). Psychiatric implications of cancer genetic testing. *Cancer, 121,* 341−360.

Holm, I. A., Savage, S. K., Green, R. C., Juengst, E., McGuire, A., Kornetsky, S., ... Taylor, P. (2014). Guidelines for return of research results from pediatric genomic studies: Deliberations of the Boston Children's Hospital Gene Partnership Informed Cohort Oversight Board. *Genetics in Medicine, 16,* 547−552.

Hufnagel, S. B., Martin, L. J., Cassedy, A., Hopkin, R. J., & Antommaria, A. H. M. (2016). Adolescents' preferences regarding disclosure of incidental findings in genomic sequencing that are not medically actionable in childhood. *American Journal of Medical Genetics Part A, 170,* 2083−2088.

Jarvik, G. P., Amendola, L. M., Berg, J. S., Brothers, K., Clayton, E. W., Chung, W., ... Wolf, W. A. (2014). Return of genomic results to research participants: The floor, the ceiling, and the choices in between. *The American Journal of Human Genetics, 94,* 818−826.

Kalia, S. S., Adelman, K., Bale, S. J., Chung, W. K., Eng, C., Evans, J. P., ... Miller, D. T. (2017). Recommendations for reporting of secondary findings in clinical exome and genome sequencing, 2016 update (ACMG SFv2. 0): A policy statement of the American College of Medical Genetics and Genomics. *Genetics in Medicine, 19*(249).

Kleiderman, E., Knoppers, B. M., Fernandez, C. V., Boycott, K. M., Ouellette, G., Wong-Rieger, D., ... Avard, D. (2014). Returning incidental findings from genetic research to children: Views of parents of children affected by rare diseases. *Journal of Medical Ethics, 40,* 691−696.

Levenseller, B. L., Soucier, D. J., Miller, V. A., Harris, D., Conway, L., & Bernhardt, B. A. (2014). Stakeholders' opinions on the implementation of pediatric whole exome sequencing: Implications for informed consent. *Journal of Genetic Counseling, 23,* 552−565.

Mackley, M. P., Fletcher, B., Parker, M., Watkins, H., & Ormondroyd, E. (2017). Stakeholder views on secondary findings in whole-genome and whole-exome sequencing: A systematic review of quantitative and qualitative studies. *Genetics in Medicine, 19,* 283−293.

McCullough, L. B., Slashinski, M. J., McGuire, A. L., Street, R. L., Eng, C. M., Gibbs, R. A., ... Plon, S. E. (2016). Is whole-exome sequencing an ethically disruptive technology? Perspectives of pediatric oncologists and parents of pediatric patients with solid tumors. *Pediatric Blood & Cancer, 63,* 511−515.

Middleton, A., Morley, K. I., Bragin, E., Firth, H. V., Hurles, M. E., Wright, C. F., & Parker, M. (2016). Attitudes of nearly 7000 health professionals, genomic researchers and publics toward the return of incidental results from sequencing research. *European Journal of Human Genetics, 24,* 21−29.

Phillips, D. L. (1986). *Toward a just social order.* Princeton, NJ: Princeton University Press.

Rigter, T., Van Aart, C., Elting, M., Waisfisz, Q., Cornel, M., & Henneman, L. (2014). Informed consent for exome sequencing in diagnostics: Exploring first experiences and views of professionals and patients. *Clinical Genetics, 85,* 417−422.

Ruddick, W. (1979). Parents and life prospects. In O. O'Neill, & W. Ruddick (Eds.), *Having children: Philosophical and legal reflections on parenthood* (pp. 123−137). New York: Oxford University Press.

Sapp, J. C., Dong, D., Stark, C., Ivey, L. E., Hooker, G., Biesecker, L. G., & Biesecker, B. B. (2014). Parental attitudes, values, and beliefs toward the return of results from exome sequencing in children. *Clinical Genetics, 85*, 120–126.

Shahmirzadi, L., Chao, E. C., Palmaer, E., Parra, M. C., Tang, S., & Gonzalez, K. D. F. (2014). Patient decisions for disclosure of secondary findings among the first 200 individuals undergoing clinical diagnostic exome sequencing. *Genetics in Medicine, 16*, 395–399.

Tan, N., Amendola, L. M., O'Daniel, J. M., Burt, A., Horike-Pyne, M. J., Boshe, L., . . . Jarvik, G. P. (2017). Is "incidental finding" the best term? A study of patients' preferences. *Genetics in Medicine, 19*, 176–181.

Van El, C. G., Cornel, M. C., Borry, P., Hastings, R. J., Fellmann, F., Hodgson, S. V., . . . de Wert, G. M. W. R. (2013). Whole-genome sequencing in health care. *European Journal of Human Genetics, 21*, 580–584.

Van El, C. G., Dondorp, W. J., De Wert, G. M. W. R., & Cornel, M. C. (2013). Call for prudence in whole-genome testing. *Science, 341*, 958–959.

Wolf, S. M., Lawrenz, F. P., Nelson, C. A., Kahn, J. P., Cho, M. K., Clayton, E. W., . . . Wilfond, B. S. (2008). Managing incidental findings in human subjects research: Analysis and recommendations. *The Journal of Law, Medicine & Ethics, 36*, 219–248.

Wouters, R. H. P., Bijlsma, R. M., Frederix, G. W. J., Ausems, M. G. E. M., van Delden, J. J. M., Voest, E. E., & Bredenoord, A. L. (2018). Is it our duty to hunt for pathogenic mutations? *Trends in Molecular Medicine, 24*, 3–6.

Wouters, R. H. P., Cornelis, C., Newson, A. J., Bunnik, E. M., & Bredenoord, A. L. (2017). Scanning the body, sequencing the genome: Dealing with unsolicited findings. *Bioethics, 31*, 648–656.

CHAPTER 10

Genome Sequencing in Prenatal Testing and Screening: Lessons Learned From Broadening the Scope of Prenatal Genetics From Conventional Karyotyping to Whole-Genome Microarray Analysis

Sam Riedijk*, Karin Diderich, Robert-Jan Galjaard and Gosia Srebniak
Clinical Genetics, Erasmus Medical Centre, Rotterdam, The Netherlands

Whole exome sequencing (WES), that allows more detailed analysis of fetal material than chromosomal microarrays, enabled a revolution in prenatal genetic testing. The traditional phenotype-first approach has been challenged, especially in prenatal settings. As prenatal phenotyping is limited to ultrasound imaging and most of the patients are curious about the intellectual abilities of their future child the genotype-first approach is especially interesting in the context of prenatal screening (PNS) and testing. Currently, the new sequencing technology that allows rapid sequencing of the whole genome/exome enabled the following tests in prenatal context:

1. *Exome sequencing in fetal material in case of ultrasound anomalies.* When ultrasound anomalies are detected on the routine prenatal ultrasound examination chromosomal microarrays are nowadays recommended in prenatal settings. However, in most cases the cause of the fetal anomalies remains unknown. So far only targeted mutation testing was possible when fetal phenotype was suggestive for a particular genetic disorder (e.g., Noonan syndrome, Smith—Lemli—Opith syndrome). The next-generation sequencing allows broad mutation analysis even when phenotypic data are limited and offers additional detection of genetic causes of ultrasound anomalies. Recent studies showed an increased diagnostic yield of about 6%—80% depending on the referral reason and patient selection (Best et al., 2018).

2. *Cell-free fetal DNA (cfDNA) or fetal cells testing in maternal plasma.* So called noninvasive prenatal testing (NIPT) has already revolutionized PNS. Next-generation

---

* This author wrote over 90% of this chapter.

Clinical Genome Sequencing
DOI: https://doi.org/10.1016/B978-0-12-813335-4.00010-6

157

sequencing and sophisticated bioinformatics allow not only detection of chromosomal aberrations, but detection of fetal genomic mutations in maternal plasma as well. The high-throughput character of next-generation sequencing allowed overcoming of the huge problem of maternal contamination, which has always been a limitation when microarray techniques are used. However, the cfDNA that is most commonly being tested originates from placental and not directly from a fetal tissue and it may also originate from other maternal tissues, therefore it is recommended to confirm positive results in an invasively sampled fetal tissue (Dondorp et al., 2015; Srebniak et al., 2018). The phenomenon of placental mosaicism is the most common explanation for both "false" positive and "false" negative results. The cfDNA analyses are currently limited when compared to chromosomal microarray and exome sequencing in invasively obtained fetal material, but the noninvasive character of this test led to an introduction of the cytogenomic analysis in the general population (the NIPT trisomy test is now offered in all pregnancies). The genotype-first approach is therefore already widely applied in PNS setting.

The switch from the phenotype-first approach to the genotype-first approach is challenging. The new technology (both next-generation sequencing and chromosomal microarray) allows broad whole-genome testing that potentially may lead to a discovery of unexpected diagnoses. As it becomes cheaper and more accessible, the clinical implementation becomes more available. The genotype-first approach has been proven to be more powerful than our conventional phenotype–first approach followed by conventional karyotyping (CK) (Levy & Wapner, 2018; Wapner et al., 2012). This and the fact that fetal phenotyping is limited lead naturally to less stringent clinical selection of patients that are offered broader genomic tests (e.g., implementing chromosome microarray testing in pregnancies without ultrasound referred for invasive prenatal diagnosis [PND]) or even to the situation when a genetic test is offered to the general pregnant population (e.g., NIPT for aneuploidy and a selection of microdeletion syndromes). The current situation makes the boundaries between screening and diagnostics blurry. The natural mechanism is to defend the boundaries and apply new regulations that preserve the conventional settings, however we cannot forget that future parents' need for information about the child also evolves within time and new research brings new insights. For example, although new data on applying genome-wide microarray testing in pregnancies without ultrasound anomalies showed that the prevalence of early-onset syndromic disorders caused by submicroscopic chromosome aberrations is actually higher than the prevalence of Down syndrome in young women, prenatal microarray testing is still not widely accepted and performed (Srebniak et al., 2018), whereas screening for Down syndrome is recognized as a main scope of PNS.

To be able to answer the question whether the new developments are beneficial for patients in prenatal settings we have extensively investigated patients' choices and perspectives in the context of broad prenatal testing and noninvasive screening. The lessons learned from our studies can be used when exome sequencing is being implemented in clinical settings.

Various stakeholders in the field of prenatal genetic testing have been debating the implications of the implementation of fast-developing techniques. In 2012, on the verge of replacing CK by whole-genome microarray analysis, the discussions in the literature were about whether an expanded scope of prenatal genetic testing could be aligned with informed choice (De Jong et al., 2014; Dondorp, Sikkema-Raddatz, de Die-Smulders, & de Wert, 2012; Shuster, 2007). Professionals were worried about information overload that would psychologically burden the pregnant couple. Concerns were voiced that prenatal genetic testing could generate knowledge about the unborn child that would not lead to ending a pregnancy but would impede the future child's autonomy. Also, the possibility of findings that would not lead to medical action of some sort was seen as undesirable. In 2015, when NIPT made its entrance in the PNS field, the same worries were voiced by the same professionals in a joint statement by the European and American Societies of Human Genetics (ESHG and ASHG, respectively) (Dondorp et al., 2015).

While the field is struggling to implement whole-genome microarray in ways maximally beneficial to pregnant couples, a new technique has already entered the stage, and its significantly higher resolution magnifies the topics of debate (Farrell & Allyse, 2018; Narayanan, Blumberg, Clayman, Pan, & Wicklund, 2018; Richardson & Ormond, 2017; Westerfield et al., 2015). In a mere 6 years, whole-genome microarray has become an accepted technique for PND and the discussion now is about replacing microarray with WES. In order to provide some guidance, the ACMG made a position statement providing over 57 genetic conditions they advised to report back prenatally to pregnant couples (Green et al., 2013a). The ACMG stated that a variety of challenging outcomes may be generated by the even more detailed technique of prenatal WES. They proposed that pregnant couples should be made fully aware of the difficulties of these outcomes in order to provide them with opt-out opportunities (Green et al., 2013b). The ESHG took a more conservative stance and advised targeted use of WES (van El et al., 2013). In 2018, the International Society for Prenatal Diagnosis, the Society for Maternal Fetal Medicine, and the Perinatal Quality Foundation made a joint position statement about the use of genome-wide sequencing for fetal diagnosis (International Society for Prenatal Diagnosis, Society for Maternal and, Fetal Medicine, Perinatal Quality Foundation, 2018). In summary, genome-wide sequencing is viewed as acceptable in case regular techniques find no genetic causes for ultrasound anomalies. Incidental findings and the scarce information about genotype—phenotype correlations are viewed as challenging. This statement prescribes that teams work in a multidisciplinary setting, and extensive in-depth pre- and post-test counseling are offered to pregnant couples. Although the challenges accompanying prenatal WES are eloquently described, empirical research addressing these issues to find solutions is scarce.

To be able to answer the question whether the new developments are beneficial for patients in prenatal settings we have extensively investigated patients' choices and perspectives in the context of broad prenatal testing and noninvasive screening. The lessons learned from our studies can be used when next-generation sequencing is being implemented in clinical settings.

2012—A NEW ERA IN PND: DO PREGNANT COUPLES WANT A BROADER SCOPE OF INVASIVE PRENATAL GENETIC TESTING?

Interestingly, while most of the discussions were held amongst various professionals, pregnant couples were hardly given a say in what they considered a desired prenatal genetic testing offer. This made us decide to offer pregnant couples real choices about the scope of their prenatal genetic test. Since we started in 2007 using whole-genome microarray analysis in case of ultrasound fetal anomalies, we offered pregnant couples a choice regarding the extent to which they wished to be informed about predefined categories of genetic anomalies (Srebniak et al., 2011). After extensive counseling pregnant couples were asked to tick a box on their consent form. Most pregnant couples chose to be informed of any (susceptibility for) severe condition(s) of any age of onset. This led us to consider that pregnant couples may not share the concerns that professionals were voicing about them.

In 2012, we replaced CK by 0.5 Mb whole-genome microarray analysis for pregnancies without ultrasound anomalies. Between 2012 and 2013, we offered pregnant couples a choice between a single nucleotide polymorphism (SNP) array analyzed at 5 Mb resolution (more or less comparable to CK) and an SNP array analyzed at 0.5 Mb resolution (higher resolution) within a research setting. We thus offered the high-risk pregnant couples a choice to opt-out of our broad-scope prenatal genetic test. Those opting for 0.5 Mb were offered the additional choice of being informed of susceptibility loci (SL) if found. SL were explained as "risk factors"; genetic variants that give an elevated but unquantified chance on mainly neurodevelopmental disorders, such as autism, learning disabilities, epilepsy, and/or psychiatric disorders. We provided extensive pretest counseling, paying much attention to illustrating and discussing the difference between the narrower and the broader scope array. Examples of what could be detected additionally by 0.5 Mb testing over 5 Mb testing were illustrated with Wolf—Hirschhorn syndrome, Duchenne muscular dystrophy, and examples of SL.

Our policy to start offering 0.5 Mb microarray for all indications was widely criticized by other genetic centers in the Netherlands which chose to employ a technique that would only reveal the presence of the common trisomies, which was in accordance with the national PNS program (RIVM, 2018). Moreover, it was argued that it was unfair to offer a broad scope of PND to the high-risk pregnant population, whereas in the rest of the country high-risk pregnancies would only be diagnosed for trisomies 13, 18, and 21. We argued that the maintained scope of PNS, that is, to detect trisomies 13, 18, and 21, is at odds with exerting reproductive autonomy. If prospective parents want to be informed whether their child will have a severe genetic condition in order to make autonomous decisions about their pregnancy, then a targeted analysis will serve their reproductive autonomy far less than an untargeted whole-genome analysis. Similar discussions were seen in the international literature. Some were concerned that testing outside the scope of PNS might unnecessarily burden pregnant couples (de Jong et al., 2013). Others counterargued that withholding information would be paternalistic and should be avoided (McGillivray, Rosenfeld, McKinlay Gardner, & Gillam, 2012). In line with the core principles of genetic counseling we adopted the policy that all high-risk pregnant couples should have access to broad-scope prenatal diagnoses to enable exerting reproductive autonomy.

To evaluate and validate our policy we studied the acceptance of uptake of the broad prenatal test. We aimed to investigate whether pregnant couples at increased risk for an aneuploidy prefer 5 Mb (narrower scope, comparable to CK) or 0.5 Mb array (broader scope). We assessed whether couples who engage in PNS or PND differed in this choice (theoretical vs actual choice) and inquired whether pregnant couples wished to decide about the scope of prenatal testing.

Methods and procedure

Pregnant women or couples were included between February 2012 and September 2013 in the clinic prenatal medicine in the Erasmus Medical Center in the Netherlands. Consenting participants were included if they had an (1) increased risk of common trisomies [advanced maternal age, increased risk based on first trimester screening (FTS), or combined indication], (2) the woman or couple was participating in first-trimester PNS or invasive PND, and if (3) they were fluent in the Dutch language. We excluded participants in the case of ultrasound anomalies and/or language barriers. After consenting, an additional genetic counseling with a clinical geneticist or a trained research assistant by telephone was planned before the actual PNS or PND in order to enable informed decision-making. Sessions took 45 minutes on average.

During this counseling, extensive information was provided. Five Mb resolution array was presented as a "narrower test" detecting trisomies 13, 18, and 21 and other microscopically visible deviations. The additional diagnostic yield of 0.5 Mb resolution compared to 5 Mb was explained using examples such as Wolf–Hirschhorn syndrome, Duchenne muscular dystrophy, and examples of SL. We taught the participants that SL were "risk factors," that give an elevated but unquantifiable chance on mainly neurodevelopmental disorders, such as autism, learning disabilities, epilepsy, and/or psychiatric disorders. Risk factors were explained to occur in both healthy and affected individuals, with expression ranging from none to severe. During the last part of the counseling the couples' concerns and questions were addressed. Within 3 days after counseling but before their PND or PNS appointment, all participants filled out a questionnaire individually.

Participants engaging in PND were contacted by the researcher 1 day before their appointment to ascertain their choice (0.5 Mb or 5 Mb analysis). The laboratory was informed of the couples' choice and performed their array resolution of choice. Participants engaging in PNS and not proceeding with PND made a hypothetical choice. PNS participants filled out the questionnaire hypothetically.

One hundred forty-one pregnant couples participated in the study. A total of 59 couples were included in our study based on the first trimester combined test (FTCT) (van der Steen et al., 2014). A total of 82 couples with a maternal age indication first engaged in the FTCT. These couples made a hypothetical choice. As Fig. 10.1 illustrates, a vast majority of the 141 included pregnant couples chose broad-scope prenatal testing (0.5 Mb resolution array).

We observed an interesting difference between these groups. Significantly more couples in the group engaging in PND chose a broader scope (95%), including SL, compared to the group making a hypothetical choice (69%).

Reproductive autonomy was indeed highly valued by the pregnant couples. When asked, most of the couples wanted to decide about the scope of prenatal testing themselves. A minority thought the healthcare practitioner should make the decision.

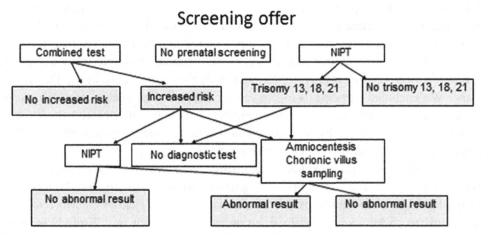

Figure 10.1 Overview of prenatal screening offer in the Netherlands.

Prenatal WES will be highly accepted among high-risk pregnant couples

Given the strong wish of pregnant couples to decide about the diagnostic scope of their prenatal genetic test and that when given a choice most pregnant couples opt for maximum information, we expect pregnant couples to have a high interest for prenatal WES. Indeed, recent studies have shown that pregnant couples looking for the cause of ultrasound anomalies were keen on receiving as much information as possible from whole-genome microarray analysis (Baker et al., 2018) and WES (Quinlan-Jones et al., 2016; Quinlan-Jones, Hillman, Kilby, & Greenfield, 2017). It can be expected that offering prenatal WES will lead to even more poignant differences in the various views on the desirability of expanding the scope of PND among health care practitioners (HCPs) (Narayanan et al., 2018). More studies are needed in which pregnant couples are offered real choices and in which the impact of prenatal WES is assessed. Furthermore, such empirical studies should inform how to care best for pregnant couples engaging in prenatal WES.

ARE PREGNANT COUPLES MAKING INFORMED DECISIONS ABOUT THE SCOPE OF THEIR PRENATAL GENETIC TEST?

Making a real-time offer about the diagnostic scope entailed the responsibility of assessing whether couples indeed made informed choices. According to the ESHG/Eurogentest a prerequisite for reproductive autonomy is making an informed choice after extensive pretest counseling. Within the above-described study, we measured informed decision-making according to the operationalization of informed choice by Marteau et al. (Marteau, Dormandy, & Michie, 2001). We assessed whether couples' choices were based on knowledge, in accordance with their personal attitudes and values, and whether their choice was behaviorally implemented, which is the eventual choice for either scope. Choices based on sufficient knowledge and in accordance with personal attitudes were defined as completely informed; choices based either on sufficient knowledge but in discordance with personal values or based on insufficient knowledge but in accordance with personal values were defined at partly informed; and choices based both on insufficient knowledge and in discordance with personal values were defined as uninformed choices.

Overall, 62.3% made a completely informed choice. A partly informed choice was made by 33.3% of women, 24.6% had poor knowledge but a consistent attitude, and 8.7% had good knowledge but an inconsistent attitude. Lastly, 4.3% made a completely uninformed choice.

In spite of all of the controversy about prenatal whole-genome microarrays, we found that the majority of pregnant women in our study were capable of making an informed choice regarding the scope of their prenatal genetic test. In addition, we found no clinically elevated levels of anxiety and anxiety decreased significantly over time, independent of which scope was chosen, as has been found previously in similar studies (van der Bij, de Weerd, Cikot, Steegers, & Braspenning, 2003). Overall, the knowledge scores of our participants, using a relatively conservative cut-off score compared to other studies, were largely sufficient. However, although most of the pregnant couples made informed choices, we were concerned. Given the extensive pretest counseling and the relatively high average educational level of our sample, we would expect most participants to have sufficient knowledge scores. Yet, we found that a substantial group of pregnant couples had difficulties in correctly answering the question of what a susceptibility locus is.

The question thus arises as to how much weight we should attach to the level and details of knowledge. We hypothesized that we might be seeing individual differences with regard to informational needs and the level of detail people feel comfortable with for decision-making, as seen in genetic counseling for hereditary cancer syndromes (Vos et al., 2013). Indeed, two types of decision-making have been described: maximizers and satisficers (Schwartz et al., 2002). Maximizers need much more information than satisficers, basing their decisions on better research and better arguments. However, satisficers are more content with their decision. Translating to prenatal decisions, generic information that testing may provide "information about severe, incurable conditions" may be adequate for satisficers, whereas maximizers need to know what conditions exactly are included in the test, in order to make an informed decision. Translating to clinical practice this implies that pretest counseling should be offered in layers: offering basic, essential information to all pregnant couples, and entering more into details upon the pregnant couple's request (Bunnik, Janssens, & Schermer, 2013). Back to our findings, if the extent of desired knowledge varies dependent upon informational needs, then it makes sense that knowledge scores will vary as well, independent of educational level, and it may not be necessary to maintain a conservative cut-off score to determine what is sufficient knowledge.

Nine percent of our pregnant couples made a choice based on sufficient knowledge, but inconsistent with their attitude. In our view, this is more problematic since attitude consistency is not a spectrum but a necessary condition for informed choice. Efforts aimed at improving informed choices mostly target the knowledge component (Schoonen, Essink-Bot, et al., 2010, Schoonen, van Agt, et al., 2010, 2012). However, making an attitude-inconsistent choice seems to imply that one's choice does not match what a person wants (and who he/she is). This, to us, seemed to be a more worrisome variant of uninformed decision-making. Sound prenatal counseling involves profoundly exploring the couple's values and attitudes to enable them to make decisions (implement behavior) that fit their personal values. Interventions aimed at improving informed choices through attitude consistency may be more effective and also more important than those targeting knowledge only (Michie, Dormandy, & Marteau, 2002; Van den Berg, Timmermans, Ten Kate, Van Vugt, & Van der Wal, 2006).

Attitude-consistent prenatal decision-making will become more prominent aspects of informed decision-making

A study measuring informed choice regarding NIPT, thus a narrow scope, found that maintaining a threshold of 75% correct answers for sufficient knowledge, only 44% scored above threshold (Piechan et al., 2016). With the increasing complexity of techniques and the increasing scope of results these techniques may generate it has become clear that it is impossible to acquire sufficient knowledge for specific outcomes. The question then becomes a more generic one: do prospective parents want to know if there is an increased chance that their child may have a genetic condition? Deliberation about what constitutes a satisfactory quality of life of one's child and about one's resources to cope with a genetic condition and about one's tolerance for uncertainty then may need to become more prominent aspects of informed decision-making.

WHAT IS THE IMPACT OF RECEIVING HIGHLY UNCERTAIN PRENATAL GENETIC TEST RESULTS?

We were interested in learning what the impact of receiving an SL would be. However, in our sample none of the pregnant couples received an SL as a prenatal test result. In our daily prenatal clinic, we implemented 0.5 Mb microarray analysis for all indications, and within the timeframe of 1.5 years (2012–13) we found an SL in 14 couples. In the pretest counseling SL were explained to the couples as giving an increased risk of variable neurodevelopmental problems which can also be found in healthy people, for example, a parent. According to our policy and informed consent form, the SL was disclosed as we considered these to be actionable. Of these couples, eight women and four of their partners agreed to be interviewed to share their experiences (Van der Steen et al., 2014). The interviewees reported that receiving an SL was initially "shocking" for five parents, while the other seven felt "worried." Werner-Lin et al. (2016) reported similar reactions. The parents were reassured by swift genetic counseling. In several cases the SL was found in one of the parents and this finding relieved the parents. Our clinical geneticists were easily accessible to these parents, offering post-test counseling quickly and as often as necessary and putting the SL into perspective. Sometimes finding the SL in one of the parents explained certain neurodevelopmental problems running in the family. A key factor in mitigating parental anxiety with SL disclosure appeared to be swift access to post-test genetic counseling. Ten out of 12 participants indicated they would like to be informed about the SL again, 2 were unsure. Most interviewees in our study had no enduring worries. However, a similar study found parents to be more preoccupied with the SL during the year after disclosure (Werner-Lin, Walser, Barg, & Bernhardt, 2017). At this point, we cannot explain these differences.

Participants unanimously indicated that pregnant couples should have an individualized pretest choice about SL (non)disclosure. Before these interviews we offered no choice but instead stated in our information leaflet and informed consent form that SL would be fed back if found. After this study we adapted our policy and informed consent regarding the disclosure of SL. We started offering the possibility of receiving SL as part of our pretest counseling.

We expect a high tolerance for uncertainty and we foresee a high need for swift and easily accessible post-test counseling in prenatal WES

A 2015 study among 494 adults participating in an NIH genome sequencing study demonstrated that participants with lower tolerance for uncertainty and lower dispositional optimism were less interested in receiving sequencing results they perceived as ambiguous (i.e., lacking reliability, credibility, or adequacy) (Taber et al., 2015a, 2015b). Pretest counseling is a necessary condition to perceive sequencing results as clear or ambiguous. It is the question whether dispositional optimism and tolerance for uncertainty will similarly moderate the wish to learn ambiguous results about one's pregnancy. A study into the experiences of pregnant couples with WES for the ultrasound anomalies found during their pregnancy reported that the couples struggled with the clinical uncertainty relating to the cause and prognosis (Quinlan-Jones et al., 2017). When offering prenatal WES, it is likely that the number of SL-like findings will increase. We recommend that these couples are offered post-test genetic counseling by a specialized clinical geneticist within 24 hours after learning the test results. In our clinic, we have learned that patients particularly appreciate it when the clinical geneticist is easily accessible to them, in person or by email/phone. Having to wait a whole weekend with many questions may be unbearable and therefore we would suggest, very practically, disclosing results on Monday and offering post-test counseling on Tuesday, leaving time to digest and ask more questions during the remainder of the week. When a finding is explained clearly and put in perspective most pregnant couples that we see are relieved. Given this experience, we expect pregnant couples engaging in prenatal WES to display a high degree of tolerance for uncertain outcomes if firmly embedded in post-test counseling. Sensitive and thorough counseling will enable pregnant couples to deal with findings from WES (Mackie, Carss, Hillman, Hurles, & Kilby, 2014).

2014: A NEW ERA IN PNS: THE INTRODUCTION OF NIPT IN THE NETHERLANDS

When NIPT was introduced in the Netherlands in 2014, it was offered only to high-risk pregnant couples as an extra step between the combined test and invasive testing by chorionic villus sampling or amniocentesis. In parallel with the national implementation of NIPT we conducted a study to assess the choices, informed decisions, and psychological impact of offering the choice between NIPT and invasive prenatal testing. We employed 0.5 Mb microarray for analyzing the genome when invasive testing was performed. Other Dutch academic centers, however, used rapid aneuploidy detection. Thus, in all centers invasive testing had scope similar to NIPT, whereas in our center invasive testing offered a much broader scope of information about the fetus than NIPT. The absence of a miscarriage risk was an advantage of NIPT, but NIPT entailed significantly narrowing the scope compared to invasive testing. We found striking differences in the choices couples made compared to the rest of the Netherlands. At a national level, almost 97% chose NIPT and less than 3% of high-risk pregnant couples chose invasive testing (van Schendel et al., 2016). Among the 181 pregnant women in our study, more than 18% chose invasive testing, versus 80% NIPT (van der Steen et al., 2018).

Interestingly, when we measured whether pregnant women were making informed decisions we found a high level of attitude inconsistency (van der Steen et al., 2018). As much as 87% of the participants choosing NIPT wanted more information about other (sub)chromosomal abnormalities than NIPT can offer. We found that the pregnant women placed a much higher priority on not being at risk of a miscarriage, even if that meant potentially missing genetic information about their fetus that they indicated to consider as important. This weighing of limitations and benefits of both tests illustrates a thorough pretest counseling process, which is important in order to be able to disclose possibly unexpected results (Levy & Wapner, 2018).

Supporting the notion that these pregnant couples wanted more information about their fetus was found when we interviewed 17 women and 12 of their partners who received an incidental finding from NIPT. All experienced shock when they received this result. However, over 90% indicated they believed that incidental findings should always be communicated, so that pregnant couples could make their own reproductive decisions (Bakkeren et al., 2017).

In this study, a clear majority of the pregnant couples indicated that they wanted to make an autonomous choice about the scope of their prenatal genetic test. Our finding that women had a high interest in a broader scope of prenatal testing, and felt it was highly desirable to be informed of incidental findings, is consistent with other research (Farrell, Agatisa, & Nutter, 2014; Hill, Fisher, Chitty, & Morris, 2012; Van der Steen et al., 2014; van Schendel et al., 2014).

2017: NIPT FOR ALL PREGNANT WOMEN IN THE NETHERLANDS

Since April 2017 NIPT has been offered to all pregnant women in the Netherlands (see Fig. 10.1). This means that NIPT has become an alternative to the FTCT, and NIPT is still an option for couples with an increased risk found with the combined test. Before NIPT was introduced in the Netherlands, around 24% of all pregnant couples opted for the combined test, which was a much lower uptake that in most European countries (Bakker, Birnie, Pajkrt, Bilardo, & Snijders, 2012; Fransen et al., 2010). Currently, only a few pregnant couples still opt for the combined test, around 5%. Altogether, the uptake of first-trimester PNS seems to have doubled since NIPT was introduced for high-risk and low-risk pregnant couples. Nowadays, around 45% of the low-risk pregnant couples in the Netherlands opt for NIPT (KD, personal communication) and they are offered a choice of whether or not they wish to be informed of incidental findings; as much as 80% of the low-risk pregnant couples do (personal communication).

Prenatal WES will be highly accepted among high-risk pregnant couples

Our study showed that significantly more pregnant couples accept a miscarriage risk for a broad diagnostic scope of invasive testing than for a scope similar to NIPT. This demonstrated that pregnant couples valued the potential of learning more about the health of their fetus. Also, a vast majority of low-risk pregnant couples engaging in NIPT wish to be informed of findings other than trisomies 13, 18, and 21. At one point WES offered through NIPT, noninvasive prenatal WES (NIPW), might expand the diagnostic scope even further. However, first we would offer WES to pregnancies complicated by ultrasound abnormalities. Preliminary data in our center show that prenatal WES has an additional diagnostic yield of 20% over whole-genome microarray analysis (G. Srebniak, personal communication). A literature review on the diagnostic yield of prenatal WES shows similar results with diagnostic yields varying between 10 and 57, based on nature of selected fetal malformations (Alamillo et al., 2015; Drury et al., 2015; Hillman et al., 2015). Especially in cases with a single abnormality and much uncertainty about the absence or presence of an underlying condition, a diagnosis can aid pregnant couples in their decision. At this point we notice that many couples do not choose WES in search of a diagnosis of the congenital abnormalities because of the long turnaround time, especially after targeted DNA analysis and gene panel analysis. Besides the severe conditions for which couples could opt for termination of pregnancy, it may be equally relevant to pregnant couples to find a milder condition (e.g., without intellectual disability). Such a prenatal test result may provide the relief that aids the couple to decide to continue their pregnancy.

We anticipate that offering WES prenatally will entail that more pregnant couples with a high risk for fetal aneuploidies will be interested in engaging in prenatal WES. If WES were offered through NIPT, NIPW, which is a likely scenario (Kitzman et al., 2012), then the biggest barrier to more knowledge would be taken away. We expect that in that case a significant proportion of pregnant couples will opt for this chance to receive as much information as possible.

MEASURING INFORMED DECISION-MAKING IN NONINVASIVE VS INVASIVE PRENATAL TESTING

Several authors agree that a good decision as well as good decision-making involves knowledge (Elwyn & Miron-Shatz, 2010; Marteau & Dormandy, 2001). During our study we encountered problems with measuring informed decision-making. Using the definition of Marteau et al. that an informed decision is one based on sufficient knowledge and congruent with one's attitude, we had problems with both dimensions. First, what constitutes sufficient knowledge? Should we be scoring a percentage of correct answers and then choose a cut-off? Is sufficient knowledge for a couple necessary to make a decision that is satisfactory in the long term? Since we opened the gate to much more genetic knowledge, we decided to maintain a strict cut-off score of >80% correct answers. This made us more conservative than others in the field. We did so because we wanted to maintain a critical stance toward what we were offering to pregnant couples. Wanting them to reproduce information correctly obliged us to provide extensive pretest counseling. But our problem with the knowledge dimension was not limited to the cut-off score for sufficiency.

We also needed to decide about the content of the knowledge. Other researchers in the field focused on the characteristics of the test instead of the characteristics of the conditions being tested. However, during our pretest counseling we invited people to envision the life of their child with varying conditions and to reflect upon quality of life, burden of care, resources for coping, etc., and to reflect on the possibility of ending a pregnancy and the imagined psychological impact of such a decision. During pretest counseling pregnant couples were challenged not to base their decision on the test characteristics but to focus on the outcome of their deliberation process. Knowledge then is needed about how a certain condition can affect health, quality of life, and independence. This is problematic. To oversee the ramifications of one genetic condition is already a big psychological challenge. But for a range of possible genetic conditions, it is impossible to acquire sufficient knowledge to be doing such a deliberation process. So, what then is the value of "knowledge"?

The attitude dimension of informed decision-making was also problematic. In the form we used, the attitude items were not weighted. However, some items may weigh more or less heavily to a pregnant woman in her decision process. We measured attitude as if it were unidimensional, but, our qualitative data show that attitude is rather dynamic and multidimensional. For instance, pregnant women indicated that they valued more knowledge about a prenatal test, but weighed the miscarriage risk heavier, as they conveyed to an open question in our questionnaire.

It seemed that to solve the problems with measuring informed decision-making we needed to reconsider its concept entirely.

Informed consent, informed choice, and informed decision-making having their origins in different fields and theories and measure different phenomena, even though in the literature these are often treated as synonymous. Informed consent is a procedure to obtain consent from the patient in a medical context, or as a legal principle to ensure individual autonomy (Beauchamp & Childress, 2001). Measures of informed choice focus on the outcome, based on knowledge and congruent with attitude, to assess the quality of the decision. Thus, informed choice measures tell if the decision is rational, but not whether the decision is reasoned. Informed decision-making is based on the consumer choice model of professional—patient interaction by Charles et al. (Charles, Gafni, & Whelan, 1997). According to the informed decision-making model, informed decision-making is appropriate in case of a treatment choice where there is little medical justification for one treatment option over another. This is seen in decisions made about PNS, which depend largely on personal values and preferences (Charles, Gafni, & Whelan, 1999). Informed decision-making might be a more effective way to evaluate the decision-making process in PNS, because it not only assesses knowledge and attitude, but also evaluates if this information is used to make the decision. Consequently, informed decision-making includes deliberation, which enables researchers to decide if the decision was reasoned.

We have started to develop a measure of informed decision-making for low-risk pregnant couples who choose between the combined test, NIPT, or no test. We have maintained a knowledge scale, but one that focuses on the disease characteristics of trisomies 13, 18, and 21. Moreover, instead of measuring attitude we measure deliberation. We defined deliberation as the attribution of personal meaning to the provided information and as the weighing and considering of prospective parents to what they consider to be a worthy life for their child, what termination of a pregnancy would mean to them, or what living with a child who needs more care would require of them.

With our instrument we measure whether pregnant couples have sufficient knowledge of the aneuploidies, and whether their deliberation process was sufficiently extensive. This measure thus evaluates the decision-making process rather than the outcome.

Informed decision-making is a prerequisite for reproductive autonomy and we need to create new measures to assess informed decision-making when we start to use WES in prenatal genetic testing

What we learned is that measuring informed decision-making is difficult. Alongside the technical developments in prenatal genetics, we gathered insights that have changed our views on what constitutes adequate informed decision-making. These insights should inform the development of measures that are applicable in this quickly advancing field. When WES will be offered in prenatal genetic screening, we may have to change our ideas of what is sufficient knowledge entirely yet again (Horn & Parker, 2018).

A qualitative study among parents of children with rare diseases showed that these parents were content not receiving secondary findings from WES (Mackley et al., 2018).

Indeed, a large recent review showed that when knowledge is low, pregnant couples seem to want much knowledge from WES from a sense of "right," whereas when knowledge increased, pregnant couples displayed more caution about secondary findings (Mackley, Fletcher, Parker, Watkins, & Ormondroyd, 2017). When pregnant couples have a greater understanding of the implications and limitations of WES, they might have a better appreciation of the uncertainty they may have to deal with. From counseling using whole-genome microarray we already gained many insights into the dynamics of uncertainty (Werner-Lin, McCoyd, & Bernhardt, 2016).

During our study we shifted our focus from attitude to deliberation. Chen and Wasserman (2017) opinionated that prenatal WES should be offered without disclosure restriction, but that instead the autonomy of pregnant couples should be enhanced by stimulating reflection on values, goals, and capacities in order to decide what kind of knowledge they wish to obtain from WES (Chen & Wasserman, 2017).

THE ROLE OF THE COUNSELOR IN DECIDING ABOUT PRENATAL GENETIC TESTING

Prenatal counselors offered pregnant couples choices between follow-up prenatal genetic tests. The primary goal of such pretest counseling is to enable informed choice. Traditionally, genetic counselors have adopted a nondirective attitude, which promotes autonomy and self-directedness and aims for patients to feel competent, understood, and in control (Veach et al., 2007; Kessler, 1997; Evans, 2006). Counselors provide information about the options a pregnant couple has, but also address individual attributes such as emotions, resilience, and social support. In order to obtain the various patient-centered goals, the counselor requires a broad set of counseling skills. When NIPT was introduced as an option for high-risk pregnant couples, several senior gynecologists were offering pretest counseling in our academic center. Although couples are expected to exert autonomous decision-making, we were interested in the influence counselors have on the eventual decision. Therefore, we have studied the counselors' attitude towards counseling in general and their preferred options regarding prenatal testing. Five counselors, who counseled a total of 181 pregnant couples were included in the study. Gynecologists were interviewed regarding their pretest counseling and enquired about their counseling approach. The interviews were audiotaped and transcribed, and analyzed by five raters. Counselors' counseling approach was judged on nondirectiveness, patient-centeredness, and information-centeredness. We also asked the gynecologists to give examples of how they formulated the options to the pregnant women. Significant differences were observed between counselors in the choices pregnant couples made (van der Steen et al., 2019).

There seemed to be an influence of the counselor, because all gynecologists were instructed to offer nondirective counseling and to convey the same information about

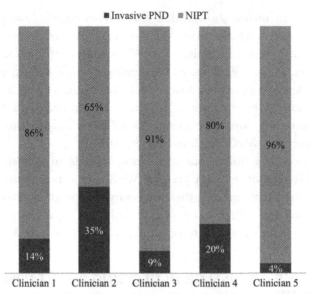

Figure 10.2 Differences in uptake of prenatal test differentiated by counselor. *From van der Steen, S. L., Houtman, D., Bakkeren, I. M., Galjaard, R. H., Polak, M. G., Busschbach, J. J., et al. (2019). Offering a choice between NIPT and invasive PND in prenatal genetic counseling: The impact of clinician characteristics on patients' test uptake. European Journal of Human Genetics, 27, 235–243.*

the different options. We could not explain the differences in choices by patient characteristics such as levels of knowledge, or differences in attitude toward prenatal testing or demographic characteristics. Nor could the differences in uptake of a specific test between counselors (see Fig. 10.2) be explained by their nondirectiveness, or by their patient- or information-centeredness. However, we found important differences in the manner in which counselors framed the information they provided. For instance, counselor number 5, who had the highest NIPT uptake said "*I say that in the experience of my patients, chorionic villus sampling is often perceived as a more unpleasant procedure*" whereas counselor 2, who had the lowest uptake of NIPT said "*I tell that with invasive PND, you test for more than just the common trisomies. We are looking at all the chromosomes on a detailed level and it is possible to detect thousands of other aberrations that might be equally relevant, and that the background risk is the same for everyone in the population. I explain them that the additional risk of a miscarriage is nihilistic.*" A significant proportion of the patients of Counselor 2 stated that they thought they made a choice that was the counselor's preference. The study offered a unique insight into the "black box" of pretest counseling during the interviews. The preliminary conclusion was that the manner in which counselors frame their information does seem to have an impact on the choices pregnant couples make. Unintentionally, framing may influence the decision-making process of patients. We concluded that in order to facilitate well-deliberated and informed choices, counselors need to reflect on their potential impact on the patients' decision-making process.

Using WES in prenatal genetic testing will likely magnify the influence of the counselor

We have observed a significant framing effect among highly experienced and skilled pretest counselors. Broadening the scope of prenatal testing with WES implies that the information that a counselor needs to frame is also widened tremendously. The magnitude of possible outcomes will only increase the impact of a counselor's framing style. We should study how we can promote autonomy and self-directedness in the era of prenatal WES. Suggested frameworks may form a basis (Chen & Wasserman, 2017), and empirical studies should follow as well.

THE PSYCHOLOGICAL IMPACT OF DECIDING ABOUT THE COURSE OF A STRONGLY DESIRED PREGNANCY

Our research projects endorsed the clinical impression that new techniques to better detect fetal genetic anomalies are highly accepted by pregnant couples. Many parents enter prenatal genetics for reassurance (Quinlan-Jones et al., 2017; Richardson & Ormond, 2017). Still, it should be noted that the couples studied were in the pretest phase and hardly any pregnant couples received an unfavorable test result.

Whilst prenatal WES offers exciting opportunities for the detection of genetic causes for fetal anomalies (Van den Veyver, 2016), the ever-increasing amount of genetic information that can be obtained also poses challenges to integrate such testing into prenatal clinical practice (Dondorp et al., 2012; Richardson & Ormond, 2017). Pregnant couples often indicate to be interested in receiving results that are uncertain (Kalynchuk, Althouse, Parker, Saller, & Rajkovic, 2015; Walser, Kellom, Palmer, & Bernhardt, 2015) but experience confusion, anxiety, and shock when they actually receive such results (Bernhardt et al., 2013; Walser, Werner-Lin, Russell, Wapner, & Bernhardt, 2016; Werner-Lin et al., 2016; Quinlan-Jones et al., 2017). It is important that we keep a clear view on the tremendous impact of having to decide about continuing or ending a deeply desired pregnancy. In this regard, HCPs display a justified careful approach when debating the possible impact of new genetic techniques on the lives of their patients.

Couples who must decide about the course of their pregnancy after receiving an unfavorable prenatal genetic test result need extensive and comprehensive counseling. Couples with more advanced pregnancies have more adjustment problems after termination of the pregnancy (Korenromp, Page-Christiaens, van den Bout, Mulder, & Visser, 2009). Also, uncertain diagnoses are associated with adjustment problems (Korenromp et al., 2009). The finding that uncertain diagnoses are related to long-term psychological adjustment problems is one that merits closer attention, because it means that an uncertain diagnosis will strain the entire psychological task of prenatal decision-making and post-decision adjustment more than a certain diagnosis would (Han, Klein, & Arora, 2011; Schoeffel, McCarthy Veach, Rubin, & LeRoy, 2018).

We wish to briefly describe these psychological processes. Pregnant couples must reach a decision in which their various thoughts, feelings, and moral considerations are reconciled.

In the literature, not much can be found on the psychological challenges parents face when deciding to continue their pregnancy after fetal anomalies have been found. Our clinical experience is that couples, particularly the pregnant woman, may have difficulties re-attaching to their child. Feelings of guilt about having considered ending the pregnancy are often described as a barrier to re-attachment.

Psychologically processing the decision to end the pregnancy is a demanding task. Leon (1990) has written extensively about the psychological impact of perinatal loss, in particular loss due to late pregnancy termination (Leon, 1990). He described four dimensions of psychological impact, two of which can be structurally recognized in the clinic: object loss (grief) and narcissistic damage (feelings of inadequacy/damaged self-esteem). He furthermore recognized the psychological damage other children in the family experienced due to the perinatal loss experience of the parents. Parents often voice concerns about the impact of their mourning process on the other children in the family. Korenromp et al. described the relatively large subgroup of women still experiencing trauma from the pregnancy termination up to 7 years afterwards (Korenromp et al., 2009). Geerinck-Vercammen et al. described people's experience of the inadequacy of social support and the heavy pressure on the partner relationship (Geerinck-Vercammen & Duijvestijn, 2004; Geerinck-Vercammen & Kanhai, 2003). Most researchers furthermore agree that being responsible for deciding about the course of pregnancy is burdening (Bijma, Van der Heide, & Wildschut, 2008; Chandler & Smith, 1998; Sandelowski & Barroso, 2005). In the clinic we have noticed that many couples are highly apprehensive of the delivery. However, afterwards many indicate their experience was ambiguous; sadness for the loss but also overwhelming love and pride for their child. Once the child is born parents often recognize familial similarities, promoting the attachment to their child, and making their grief more tangible. After the funeral or cremation, the grief process will usually start within a few weeks. A time frame in which the social surroundings of the couple return to their daily lives and employers expects the couple to return to work. Moreover, couples are challenged to stay connected with each other during this process. Gender differences have been found in long-term follow-up studies. These differences could be more appropriately renamed as pregnant/nonpregnant differences, since similar dynamics can be observed within lesbian couples. The (formerly) pregnant woman is more strongly attached to the unborn child and has a significantly more intense grieving process than the partner. The partner, in turn, takes on the task of caring for the grieving woman, the other children in the family, and the household.

Couples need to discover that they represent different paragraphs in the same chapter, and this also requires attention and psychological flexibility.

During this period, couples need to come to terms with their decision. Oftentimes, couples are still digesting all the genetic information they have received and need to attribute personal meaning to it. Before and after decision-making, couples rely heavily on their prenatal counselors (gynecologist, clinical geneticist) to give them the test results, to help them interpret the results, and be given time and space in the counseling process to translate this genetic information into personally meaningful information. Several studies have found that HCPs are uncomfortable with providing uncertain results (Bernhardt, Kellom, Barbarese, Faucett, & Wapner, 2014; Mikhaelian, Veach, MacFarlane, LeRoy, & Bower, 2013; Quinlan-Jones et al., 2016; Walser et al., 2015). Various studies propose that the problem of diagnostic uncertainty should be handled during pre- and post-test counseling (Bernhardt et al., 2014; Westerfield, Darilek, & van den Veyver, 2014). It is the question whether HCPs can provide the much-needed support if they themselves are not comfortable with uncertainty. The influence of tolerance for uncertainty, both among pregnant couples as well as HCP's needs to be further studied. A promising taxonomy has been published, addressing the various uncertainties that HCPs and patients may have to deal with in clinical genome sequencing (Han et al., 2017). The framing effect we observed among our prenatal counselors influenced pregnant couples in their choices and might be an expression of HCP's tolerance for uncertainty.

SUMMARY AND CONCLUSION

Our studies on informed decision-making about prenatal whole-genome microarray analysis showed that pregnant couples greatly valued their reproductive autonomy. Couples consistently voiced the wish to be offered the opportunity to make their own decision about the scope of their prenatal genetic test. Furthermore, we found that a significant proportion of pregnant couples chose a broader scope of prenatal genetic testing when offered the choice. We therefore expect that pregnant couples will be interested in prenatal WES for screening or testing purposes. Reproductive autonomy is served by offering pregnant couples access to techniques with the highest diagnostic yield, such as WES.

We have experienced that when firmly embedded in pretest and post-test counseling, pregnant couples could handle test results of uncertain meaning such as SL. Thus, when offering prenatal WES, care should be organized such that swift access to post-test counseling is offered. In our experience this is an important tool for mitigating distress. What should be further developed once WES is employed is how tolerance for uncertainty may be improved, in pregnant couples and possibly also in HCPs.

We learned during our study that our current tools for measuring informed decision-making, the essential requirement of PNS and testing, are inadequate. Developing a tool to measure whether pregnant couples meet the requirement of making informed decisions about prenatal WES is thus of utmost importance.

WES is likely to generate uncertain test results, which weigh heavily on decision-making. Pregnant couples often need support in making a best-fitting decision about the course of their pregnancy. In our experience the need for support increases when test results are more uncertain. Offering WES should thus be accompanied by post-test decision-making support. After the decision is made, pregnant couples may either need to re-attach to their unborn child or grieve for their loss. Ambiguous test results represent an additional weight to the psychological task of adapting. The availability of aftercare, focusing on grief counseling with special attention for the couple dynamics, may therefore be even more important when using WES.

ACKNOWLEDGMENTS

We wish to acknowledge the work of Sanne van der Steen and Iris Jansen-Bakkeren as PhD students on the studies presented in this chapter.

REFERENCES

Alamillo, C. L., Powis, Z., Farwell, K., Shahmirzadi, L., Weltmer, E. C., Turocy, J., et al. (2015). Exome sequencing positively identified relevant alterations in more than half of cases with an indication of prenatal ultrasound anomalies. *Prenatal Diagnosis, 35*(11), 1073−1078.

Baker, J., Shuman, C., Chitayat, D., Wasim, S., Okun, N., Keunen, J., et al. (2018). Informed decision-making in the context of prenatal chromosomal microarray. *Journal of Genetic Counseling, 27*(5), 1130−1147.

Bakker, M., Birnie, E., Pajkrt, E., Bilardo, C. M., & Snijders, R. J. (2012). Low uptake of the combined test in the Netherlands—Which factors contribute? *Prenatal Diagnosis, 32*(13), 1305−1312.

Bakkeren, I. M., van der Steen, S. L., Govaerts, L. C. P., van Hall, H., Feenstra, I., van Maarle, M. C., et al. (2017). *Non-invasive prenatal test (NIPT): A national study on receiving incidental findings.* Kopenhagen, Denmark: Poster session presented at the European Society of Human Genetics. Available online at http://www.abstractsonline.com/Plan/ViewAbstract.aspx?sKey = 0829594c-ed56-4ead-9bc4-95bb70edfc53&cKey = cdbc8d5a-f827-46be-9efe-548a76b82aef&mKey = %7b15A3630E-7769-4D64-A80A-47F190AC2F4F%7d.

Beauchamp, T. L., & Childress, J. F. (2001). *Principles of biomedical ethics.* Oxford, NY: Oxford University Press.

Bernhardt, B. A., Kellom, K., Barbarese, A., Faucett, W. A., & Wapner, R. J. (2014). An exploration of genetic counselors' needs and experiences with prenatal chromosomal microarray testing. *Journal of Genetic Counseling, 23*(6), 938−947.

Bernhardt, B. A., Soucier, D., Hanson, K., Savage, M. S., Jackson, L., & Wapner, R. J. (2013). Women's experiences receiving abnormal prenatal chromosomal microarray testing results. *Genetics in Medicine, 15*(2), 139−145.

Best, S., Wou, K., Vora, N., Van der Veyver, I. B., Wapner, R., & Chitty, L. S. (2018). Promises, pitfalls and practicalities of prenatal whole exome sequencing. *Prenatal Diagnosis, 38*(1), 10−19.

Bijma, H. H., Van der Heide, A., & Wildschut, H. I. J. (2008). Decision-making after ultrasound diagnosis of fetal abnormality. *Reproductive Health Matters, 16*(31S), 82−89.

Bunnik, E. M., Janssens, A. C., & Schermer, M. H. (2013). A tiered-layered-staged model for informed consent in personal genome testing. *European Journal of Human Genetics, 21,* 596—601.

Chandler, M., & Smith, A. (1998). Prenatal screening and women's perception of infant disability: A Sophie's Choice for every mother. *Nursing Inquiry, 5,* 71—76.

Charles, C., Gafni, A., & Whelan, T. (1997). Shared decision-making in the medical encounter: What does it mean? (or it takes at least two to tango). *Social Science & Medicine, 44*(5), 681—692.

Charles, C., Gafni, A., & Whelan, T. (1999). Decision-making in the physician-patient encounter: Revisiting the shared treatment decision-making model. *Social Science & Medicine, 49*(5), 651—661.

Chen, S. C., & Wasserman, D. T. (2017). A framework for unrestricted prenatal whole-genome sequencing: Respecting and enhancing the autonomy of prospective parents. *The American Journal of Bioethics, 17,* 3—18.

de Jong, A., Dondorp, W. J., Krumeich, A., Boonekamp, J., van Lith, J. M., & de Wert, G. M. (2013). The scope of prenatal diagnosis for women at increased risk for aneuploidies: Views and preferences of professionals and potential users. *Journal of Community Genetics, 4*(1), 125—135.

De Jong, A., Dondorp, W. J., Macville, M. V., De Die-Smulders, C. E., Van Lith, J. M., & De Wert, G. M. (2014). Microarrays as a diagnostic tool in prenatal screening strategies: Ethical reflection. *Human Genetics, 133,* 163—172.

Dondorp, W., de Wert, G., Bombard, Y., Bianchi, D. W., Bergmann, C., Borry, P., et al. (2015). Non-invasive prenatal testing for aneuploidy and beyond: Challenges of responsible innovation in prenatal screening. *European Journal Of Human Genetics, 23,* 1438.

Dondorp, W., Sikkema-Raddatz, B., de Die-Smulders, C., & de Wert, G. (2012). Arrays in postnatal and prenatal diagnosis: An exploration of the ethics of consent. *Human Mutation, 33*(6), 916—922.

Drury, S., Williams, H., Trump, N., Boustred, C., Gosgene., Lench, N., et al. (2015). Exome sequencing for prenatal diagnosis of fetuses with sonographic abnormalities. *Prenatal Diagnosis, 35*(10), 1010—1017.

Elwyn, G., & Miron-Shatz, T. (2010). Deliberation before determination: The definition and evaluation of good decision making. *Health Expectations, 13*(2), 139—147.

Evans C. Genetic counselling: A psychological approach. New York, USA: Cambridge University Press; 2006.

Farrell, R. M., Agatisa, P. K., & Nutter, B. (2014). What women want: Lead considerations for current and future applications of noninvasive prenatal testing in prenatal care. *Birth, 41*(3), 276—282.

Farrell, R. M., & Allyse, M. A. (2018). Key ethical issues in prenatal genetics: An overview. *Obstetrics and Gynecology Clinics of North America, 45*(1), 127—141.

Fransen, M. P., Wildschut, H. I., Mackenbach, J. P., Steegers, E. A., Galjaard, R. J., & Essink-Bot, M. L. (2010). Ethnic and socio-economic differences in uptake of prenatal diagnostic tests for Down's syndrome. *European Journal of Obstetrics & Gynecology and Reproductive Biology, 151*(2), 158—162.

Geerinck-Vercammen, C. R., & Duijvestijn, M. J. M. (2004). Rouwverwerking rond perinatale sterfte: Een veelvormig en natuurlijk proces. *Nederlands Tijdschrift voor Geneeskunde, 148,* 1231—1234.

Geerinck-Vercammen, C. R., & Kanhai, H. H. H. (2003). Coping with termination of pregnancy for fetal abnormality in a supportive environment. *Prenatal Diagnosis, 23*(7), 543—548.

Green, R. C., Berg, J. S., Grody, W. W., Kalia, S. S., Korf, B. R., Martin, C. L., et al. (2013a). ACMG recommendations for reporting of incidental findings in clinical exome and genome sequencing. *Genetics in Medicine, 15,* 565—574.

Green, R. C., Berg, J. S., Grody, W. W., Kalia, S. S., Korf, B. R., Martin, C. L., et al. (2013b). ACMG recommendations for reporting of incidental findings in clinical exome and genome sequencing. *American College of Medical Genetics and Genomics, 1,* 1—29.

Han, P. K. J., Klein, W. M. P., & Arora, N. K. (2011). Varieties of uncertainty in health care: A conceptual taxonomy. *Medical Decision Making: An International Journal of the Society for Medical Decision Making, 31*(6), 828—838.

Han, P. K. J., Umstead, K. L., Green, R. C., Joffe, S., Koenig, B., Krantz, I., et al. (2017). A taxonomy of medical uncertainties in clinical genome sequencing. *Genetics in Medicine, 19,* 918—925.

Hill, M., Fisher, J., Chitty, L. S., & Morris, S. (2012). Women's and health professionals' preferences for prenatal tests for Down syndrome: A discrete choice experiment to contrast noninvasive prenatal diagnosis with current invasive tests. *Genetics in Medicine, 14,* 905—913.

Hillman, S. C., Willams, D., Carss, K. J., McMullan, D. J., Hurles, M. E., & Kilby, M. D. (2015). Prenatal exome sequencing for fetuses with structural abnormalities: The next step. *Ultrasound Obstet Gynecol, 45*(1), 4−9.

Horn, R., & Parker, M. (2018). Opening Pandora's box?: Ethical issues in prenatal whole genome and exome sequencing. *Prenatal Diagnosis, 38*(1), 20−25.

International Society for Prenatal Diagnosis., Society for Maternal and Fetal Medicine., & Perinatal Quality Foundation. (2018). Joint Position Statement from the International Society for Prenatal Diagnosis (ISPD), the Society for Maternal Fetal Medicine (SMFM), and the Perinatal Quality Foundation (PQF) on the use of genome-wide sequencing for fetal diagnosis. *Prenatal Diagnosis, 38* (1), 6−9.

Kalynchuk, E., Althouse, A., Parker, L. S., Saller, D. N., & Rajkovic, A. (2015). Prenatal whole exome sequencing: Parental attitudes. *Prenatal Diagnosis, 35*, 1030−1036.

Kessler, S. (1997). Psychological aspects of genetic counseling. XI. Nondirectiveness revisted. *Am J Med Genet, 72*, 164−171.

Kitzman, J. O., Snyder, M. W., Ventura, M., Lewis, A. P., Qiu, R., Simmons, L. E., et al. (2012). Noninvasive whole-genome sequencing of a human fetus. *Science Translational Medicine, 4*(137), 137−176.

Korenromp, M., Page-Christiaens, G. C. M. L., van den Bout, J., Mulder, E. J. H., & Visser, G. H. A. (2009). Adjustment to termination of pregnancy for fetal anomaly: A longitudinal study in women at 4, 8, and 16 months. *American Journal of Obstetrics and Gynecology, 201*, 160.e1−160.e7.

Leon, I. G. (1990). *When a baby dies; psychotherapy for pregnancy and newborn loss.* New Haven: Yale University Press.

Levy, B., & Wapner, R. (2018). Prenatal diagnosis by chromosomal microarray analysis. *Fertility and Sterility, 109*(2), 201−212.

Mackie, F. L., Carss, K. J., Hillman, S. C., Hurles, M. E., & Kilby, M. D. (2014). Exome sequencing in fetuses with structural malformations. *Journal of Clinical Medicine, 3*, 747−762.

Mackley, M. P., Blair, E., Parker, M., Taylor, J. C., Watkins, H., & Ormondroyd, E. (2018). Views of rare disease participants in a UK whole-genome sequencing study towards secondary findings: A qualitative study. *European Journal of Human Genetics, 26*, 652−659.

Mackley, M. P., Fletcher, B., Parker, M., Watkins, H., & Ormondroyd, E. (2017). Stakeholder views on secondary findings in whole-genome and whole-exome sequencing: A systematic review of quantitative and qualitative studies. *Genetics in Medicine, 19*, 283−293.

Marteau, T. M., & Dormandy, E. (2001). Facilitating informed choice in prenatal testing: How well are we doing? *Am J Med Genet, 106*(3), 185−190.

Marteau, T. M., Dormandy, E., & Michie, S. (2001). A measure of informed choice. *Health Expectations, 4*(2), 99−108.

McGillivray, G., Rosenfeld, J. A., McKinlay Gardner, R. J., & Gillam, L. H. (2012). Genetic counselling and ethical issues with chromosome microarray analysis in prenatal testing. *Prenatal Diagnosis, 32*(4), 389−395.

Michie, S., Dormandy, E., & Marteau, T. M. (2002). The multi-dimensional measure of informed choice: A validation study. *Patient Education and Counseling, 48*, 87−91.

Mikhaelian, M., Veach, P. M., MacFarlane, I., LeRoy, B. S., & Bower, M. (2013). Prenatal chromosomal microarray analysis: A survey of prenatal genetic counselors' experiences and attitudes. *Prenatal Diagnosis, 33*(4), 371−377.

Narayanan, S., Blumberg, B., Clayman, M. L., Pan, V., & Wicklund, C. (2018). Exploring the issues surrounding clinical exome sequencing in the prenatal setting. *Journal of Genetic Counseling, 27*(5), 1228−1237.

Piechan, J. L., Hines, K. A., Koller, D. L., Stone, K., Quaid, K., Torres-Martinez, W., et al. (2016). NIPT and Informed Consent: An assessment of patient understanding of a negative NIPT result. *Journal of Genetic Counseling, 25*(5), 1127−1137.

Quinlan-Jones, E., Hillman, S. C., Kilby, M. D., & Greenfield, S. M. (2017). Parental experiences of prenatal whole exome sequencing (WES) in cases of ultrasound diagnosed fetal structural anomaly. *Prenatal Diagnosis, 37*, 1225−1231.

Quinlan-Jones, E., Kilby, M. D., Greenfield, S., Parker, M., McMullan, D., Hurles, M. E., et al. (2016). Prenatal whole exome sequencing: The views of clinicians, scientists, genetic counsellors and patient representatives. *Prenatal Diagnosis, 36*(10), 935—941.

Richardson, A., & Ormond, K. E. (2018). Ethical considerations in prenatal testing: Genomic testing and medical uncertainty. *Seminars in Fetal & Neonatal Medicine, 23*, 1—6.

RIVM. Prenatale screening op down-, edwards- en patausyndroom en SEO: Rijksinstituut voor Volksgezondheid en Milieu; 2018. Available at: https://www.rivm.nl/Onderwerpen/P/down_edwards_patau_en_SEO.

Sandelowski, M., & Barroso, J. (2005). The travesty of choosing after positive prenatal diagnosis. *Journal of Obstetric Gynecologic & Neonatal Nursing, 34*, 307—318.

Schoeffel, K., McCarthy Veach, P., Rubin, K., & LeRoy, B. (2018). Managing couple conflict during prenatal counseling sessions: An investigation of genetic counselor experiences and perceptions. *Journal of Genetic Counseling, 27*, 1275—1290.

Schoonen, H. M., Essink-Bot, M. L., Van Agt, H. M., Wildschut, H. I., Steegers, E. A., & De Koning, H. J. (2010). Informed decision-making about the fetal anomaly scan: What knowledge is relevant? *Ultrasound Obstet Gynecol, 37*, 649—657.

Schoonen, H. M., van Agt, H. M., Essink-Bot, M. L., Wildschut, H. I. J., Steegers, E. A., & De Koning, H. J. (2010). Informed decision-making in prenatal screening for Down's syndrome: What knowledge is relevant? *Patient Education and Counseling, 84*, 265—270, [Epub ahead of print].

Schoonen, M., Wildschut, H., Essink-Bot, M. L., Peters, I., Steegers, E., & de Koning, H. (2012). The provision of information and informed decision-making on prenatal screening for Down syndrome: A questionnaire- and register-based survey in a non-selected population. *Patient Education and Counseling, 87*(3), 351—359.

Schwartz, B., Ward, A., Monterosso, J., Lyubomirsky, S., White, K., & Lehman, D. R. (2002). Maximizing versus satisficing: Happiness is a matter of choice. *Journal of Personality and Social Psychology, 83*(5), 1178—1197.

Shuster, E. (2007). Microarray genetic screening: A prenatal roadblock for life? *The Lancet, 369*, 526—529.

Srebniak, M., Boter, M., Oudesluijs, G., Joosten, M., Govaerts, L., Van Opstal, D., et al. (2011). Application of SNP array for rapid prenatal diagnosis: Implementation, genetic counselling and diagnostic flow. *European Journal of Human Genetics, 19*, 1230—1237.

Srebniak, M. I., Joosten, M., Knapen, M., Arends, L. R., Polak, M., van Veen, S., et al. (2018). Frequency of submicroscopic chromosomal aberrations in pregnancies without increased risk for structural chromosomal aberrations: Systematic review and meta-analysis. *Ultrasound in Obstetrics & Gynecology, 51*(4), 445—452.

Taber, J. M., Klein, W. M. P., Ferrer, R. A., Han, P. K. J., Lewis, K. L., Biesecker, L. G., et al. (2015a). Perceived ambiguity as a barrier to intentions to learn genome sequencing results. *Journal of Behavioral Medicine, 38*(5), 715—726.

Taber, J. M., Klein, W. M., Ferrer, R. A., Lewis, K. L., Biesecker, L. G., & Biesecker, B. B. (2015b). Dispositional optimism and perceived risk interact to predict intentions to learn genome sequencing results. *Health Psychology, 34*, 718—728.

Van den Berg, M., Timmermans, D. R., Ten Kate, L. P., Van Vugt, J. M., & Van der Wal, G. (2006). Informed decision making in the context of prenatal screening. *Patient Education and Counseling, 63*, 110—117.

Van den Veyver, I. B. (2016). Recent advances in prenatal genetic screening and testing. *F1000 Research, 5*, 2591.

van der Bij, A. K., de Weerd, S., Cikot, R. J., Steegers, E. A., & Braspenning, J. C. (2003). Validation of the dutch short form of the state scale of the Spielberger State-Trait Anxiety Inventory: Considerations for usage in screening outcomes. *Journal of Community Genetics, 6*, 84—87.

van der Steen, S. L., Bunnik, E. M., Polak, M. G., Diderich, K. E. M., Verhagen-Visser, J., Govaerts, L. C., et al. (2018). Choosing between higher and lower resolution microarrays: Do pregnant women have sufficient knowledge to make informed choices consistent with their attitude? *Journal of Genetic Counseling, 27*, 85—94.

van der Steen, S. L., Diderich, K. E., Riedijk, S. R., Verhagen-Visser, J., Govaerts, L. C., Joosten, M., et al. (2014). Pregnant couples at increased risk for common aneuploidies choose maximal information from invasive genetic testing. *Clinical Genetics, 88*, 25–31.

van der Steen, S. L., Houtman, D., Bakkeren, I. M., Galjaard, R. H., Polak, M. G., Busschbach, J. J., et al. (2019). Offering a choice between NIPT and invasive PND in prenatal genetic counseling: The impact of clinician characteristics on patients' test uptake. *European Journal of Human Genetics, 27*, 235–243.

van der Steen, S. L., Riedijk, S., Polak, M., Bakkeren, I., Diderich, K., Knapen, M., et al. (2018). Non-invasive or invasive prenatal testing: Safety for the fetus overrules 'the need to know'. *Clinical Genetics*, Under revision.

Van der Steen, S. L., Riedijk, S. R., Verhagen-Visser, J., Govaerts, L. C. P., Van Opstal, A., De Vries, F. A. T., et al. (2014). Receiving a susceptibility locus as a result from SNP array in prenatal genetic testing: Quick recovery after initial shock. *Manuscript in preparation.*

van El, C. G., Cornel, M. C., Borry, P., Hastings, R. J., Fellmann, F., Hodgson, S. V., et al. (2013). Whole-genome sequencing in health care. Recommendations of the European Society of Human Genetics. *European Journal of Human Genetics, 21*(Suppl 1), S1–S5.

van Schendel, R. V., Kleinveld, J. H., Dondorp, W. J., Pajkrt, E., Timmermans, D. R., Holtkamp, K. C., et al. (2014). Attitudes of pregnant women and male partners towards non-invasive prenatal testing and widening the scope of prenatal screening. *European Journal of Human Genetics, 22*(12), 1345–1350.

van Schendel, R. V., Page-Christiaens, G. C., Beulen, L., Bilardo, C. M., de Boer, M. A., Coumans, A. B., et al. (2016). Trial by Dutch laboratories for evaluation of non-invasive prenatal testing. Part II-women's perspectives. *Prenatal Diagnosis, 36*(12), 1091–1098.

Veach, P. M., Bartels, D. M., & Leroy, B. S. (2007). Coming full circle: a reciprocal engagement model of genetic counselling practice. *J genet Couns, 16*, 713–728.

Vos, J., Menko, F. H., Oosterwijk, J. C., van Asperen, C. J., Stiggelbout, A. M., & Tibben, A. (2013). Genetic counseling does not fulfill the counselees' need for certainty in hereditary breast/ovarian cancer families: An explorative assessment. *Psychooncology, 22*, 1167–1176.

Walser, S. A., Kellom, K. S., Palmer, S. C., & Bernhardt, B. A. (2015). Comparing genetic counselor's and patient's perceptions of needs in prenatal chromosomal microarray testing. *Prenatal Diagnosis, 35*(9), 870–878.

Walser, S. A., Werner-Lin, A., Russell, A., Wapner, R. J., & Bernhardt, B. A. (2016). "Something extra on chromosome 5": Parents' understanding of positive prenatal chromosomal microarray analysis (CMA) results. *Journal of Genetic Counseling, 25*(5), 1116–1126.

Wapner, R. J., Martin, C. L., Levy, B., Ballif, B. C., Eng, C. M., Zachary, J. M., et al. (2012). Chromosomal microarray versus karyotyping for prenatal diagnosis. *The New England Journal of Medicine, 367*(23), 2175–2184.

Werner-Lin, A., Barg, F. K., Kellom, K. S., Stumm, K. J., Pilchman, L., Tomlinson, A. N., et al. (2016). Couple's narratives of communion and isolation following abnormal prenatal microarray testing results. *Qualitative Health Research, 26*(14), 1975–1987.

Werner-Lin, A., McCoyd, J. L., & Bernhardt, B. A. (2016). Balancing genetics (Science) and counseling (Art) in prenatal chromosomal microarray testing. *Journal of Genetic Counseling, 25*, 855–867.

Werner-Lin, A., Walser, S., Barg, F. K., & Bernhardt, B. A. (2017). "They can't find anything wrong with him, yet": Mothers' experiences of parenting an infant with a prenatally diagnosed copy number variant (CNV). *American Journal of Medical Genetics Part A, 173*, 444–451.

Westerfield, L., Darilek, S., & van den Veyver, I. B. (2014). Counseling challenges with variants of uncertain significance and incidental findings in prenatal genetic screening and diagnosis. *Journal of Clinical Medicine, 3*(3), 1018–1032.

Westerfield, L. E., Stover, S. R., Mathur, V. S., Nassef, S. A., Carter, T. G., Yang, Y., et al. (2015). Reproductive genetic counseling challenges associated with diagnostic exome sequencing in a large academic private reproductive genetic counseling practice. *Prenatal Diagnosis, 35*(10), 1022–1029.

CHAPTER 11

Clinical Genetic Testing and Counseling in Psychiatry

Christian G. Bouwkamp[1,2], Zaid Afawi[1,3] and Steven A. Kushner[1]
[1]Department of Psychiatry, Erasmus University Medical Center, Rotterdam, The Netherlands
[2]Department of Clinical Genetics, Erasmus University Medical Center, Rotterdam, The Netherlands
[3]Department of Family Medicine, Division of Community Health, Faculty of Health Sciences, Ben-Gurion University of the Negev, Beer-Sheva, Israel

CLINICAL VIGNETTE: THE IMPACT OF A POSITIVE FAMILY HISTORY OF MENTAL ILLNESS

We met Maria, the sibling in charge of arranging healthcare-related matters for her family. Maria was a friendly, calm, middle-aged woman with dark hair. In her living room, 12 family members had gathered for our visit. Maria recounted her family's history—she quit secondary school at a young age to help care for her father who had become psychiatrically ill. Later, she also assumed healthcare responsibility for six of her siblings who subsequently developed psychotic illnesses, a pattern that marked previous generations of the family. Family members recounted their histories of inpatient psychiatric hospitalizations, suicide attempts, acute care clinic visits, and mutual support. All of the affected family members were ultimately diagnosed with schizophrenia.

Amidst these challenges, Maria diligently ensured that her family members remained compliant with their prescribed medication, attended their scheduled clinic visits, maintained a reasonable diet, and avoided the perils of substance use. Despite the complexity of the situation, Maria was nevertheless able to coordinate a remarkable quality of life for her affected siblings. Of course, ensuring her siblings' quality of life came at a high cost to Maria herself. After decades of dedication to their care, Maria found herself in her fifties, without formal education or employment, and having never had a long-term romantic relationship, but with a life-long experience in family health care.

In the years following our first visit, we got to know the family better and saw many changes in their lives: their parents died and the affected siblings moved to an assisted-living facility, and several of them attempted suicide. Community mental healthcare budgets were also considerably under pressure, reducing their healthcare benefits and financial support.

Clinical Genome Sequencing
DOI: https://doi.org/10.1016/B978-0-12-813335-4.00011-8

In the affected family members, the onset of the first symptoms of psychiatric disorder was marked by a period of great doubt with regard to the origin of these symptoms, the persistence of these symptoms, and the reason they so profoundly struck this person. Ultimately, upon diagnosis, the affected individuals and their families would learn to accept that they had to cope with a chronic psychiatric disorder and that good health was no longer a guarantee. Interpersonal dynamics among close relatives and friends would change. Patients and their family members were often in doubt regarding whom they could confide about their symptoms: would others understand and not be prejudiced, afraid, and distance themselves? Very often, managing psychiatric disorder caused a strong narrowing of the activities of both the patient and their family members. This was also the case for Maria. The burden of caring for her parents and siblings together with the stigma still associated with psychiatric disorders had dramatically reduced her contact with extended family and caused friendships to fade away.

INTRODUCTION

A family history of mental illness has long been appreciated as the strongest known predictor of psychiatric illness risk (Austin & Peay, 2006). In this chapter, we will focus particularly on schizophrenia as a canonical psychiatric disorder for which there is arguably the most extensive literature among the severe psychiatric disorders, particularly with regard to heritable risk. The modern conceptualization of schizophrenia was initially classified in 1911 by Eugen Bleuler as a "group of schizophrenias" (Bleuler, 1950). Modern insights into the molecular genetic underpinnings are now bringing us back to Bleuler's original conceptualization as a group of disorders that we now together refer to using the general diagnosis of schizophrenia.

Psychiatric disorders are common and documented widely throughout recorded human history (Marneros, 2008; Nasser, 1987). Psychiatric hospitals offering treatment for the mentally ill have been around since at least the medieval period in the Arab world (Syed, 2002). Using the current psychiatric classification systems (American Psychiatric Association, 2013; WHO, 1992) an estimated 20% of all people will suffer from a psychiatric disorder at some point in their lives (Substance Abuse and Mental Health Services Administration, 2013). The lifetime prevalence of the most severe and highly heritable psychiatric disorders [schizophrenia, bipolar disorder, and autism spectrum disorders (ASD)] is roughly 2% (Perälä et al., 2007; Rice et al., 2012). Excluding the cases where a de novo mutation is causative, the strongest indicator for disease risk is a close relative with a psychiatric disorder (Kendler, McGuire, Gruenberg, & Walsh, 1994).

The focus of the field in the 19th century as well as for much of the 20th century was on the description and epidemiology of psychiatric disorders. Major contributors of the 19th century in this regard were German psychiatrist Emil Kraepelin and Swiss psychiatrist Eugen Bleuler. Kraepelin devised the term dementia praecox to describe a psychotic disorder characterized by rapid decline in cognitive integration which was later described as the *Group of Schizophrenias* by Bleuler (1950). Kraepelin devised a systematic classification of psychiatric disorders based on the clinical symptomatology. By some this system is considered the basis for the current classification systems, the *Diagnostic and Statistical Manual of Mental Disorders*, now in its fifth edition (DSM-V) and the *International Classification of Disease*, now in its tenth edition (ICD-10). In these classification systems, no relationship ought to be assumed with respect to the underlying etiology of the psychiatric disorders. Also notably, the aspect of prominent cognitive symptoms is not reflected in the diagnostic criteria. These current classification systems are based on the clinical phenomenology and to a large degree are not based on the biological etiology of disease. As a consequence, in the current classification systems there is considerable overlap between disease categories, such as schizophrenia and bipolar disorder (van Os & Kapur, 2009), as well as personality disorders (Perugi, Fornaro, & Akiskal, 2011). In addition, there is considerable heterogeneity within disease categories such as schizophrenia, due to the systems' requirements to meet at least a minimum number of symptoms out of a larger number of possible symptoms (Kendler et al., 1994). This inevitably leads to a very wide diversity of possible symptom combinations which currently mandate the same disease classification.

A BRIEF OVERVIEW OF PSYCHIATRIC GENETICS
Early insights and genetic studies in psychiatry

Kraepelin described the increased occurrence of psychiatric disorders in family members of patients. Since the publication of the work on the crossing of pea plants by Gregor Mendel, on what is now known as Mendelian genetics, statistical genetic and epidemiological studies were performed in groups of patients with psychiatric disorders (Schulze, Fangerau, & Propping, 2004). In these early stages, it was found that the concordance rate for schizophrenia of monozygotic twins was ~50%, which is still a valid number informing about a considerable genetic contribution to the risk of developing disease (Polderman et al., 2015). Through these large epidemiological studies, it has been possible to calculate the recurrence rates in, for example, schizophrenia (Table 11.1) (Austin & Peay, 2006).

Chromosomal studies

The study of large chromosomal rearrangements has been very informative in identifying associated chromosomal regions with human disease. Many of us will be familiar with the association of trisomy 21 and Down syndrome and maybe less so with Patau syndrome and Edwards syndrome associated with trisomies 13 and 18, respectively. For less-common chromosomal abnormalities the observation of co-occurrence of

Table 11.1 Recurrence rate × genetic distance for schizophrenia

Degree	Relationship to proband	Median risk (%)
First degree	Parent	5.6
	Child with one affected parent	13
	Child with two affected parents	46
	Sibling	9
	Sibling with one affected parent	16.85
	Monozygotic twin	48
	Dizygotic twin	13.5
	Half-sibling	4.6
Second degree	Grandchild	4.5
	Uncle/aunt	3
Third degree	Niece/nephew	4
	Cousin	1.75

Source: Adapted from Austin, J. C., & Peay, H. L. (2006). Applications and limitations of empiric data in provision of recurrence risks for schizophrenia: a practical review for healthcare professionals providing clinical psychiatric genetics consultations. Clinical Genetics, 70(3), 177–187.

disease could lead to hypotheses that the aberration is linked (or associated with the disease). With regard to mental illness a number of chromosomes have been described (Bouwkamp et al., 2017; MacIntyre, Blackwood, Porteous, Pickard, & Muir, 2003). An inherent difficulty to this method is the idiopathic nature of these chromosomal studies. If a single family is identified in which a chromosomal abnormality segregates with mental illness, this if not yet definitive proof of its cause.

The candidate gene approach

Through the working mechanism of psychotropic medication and other hypothesized biological deficiencies thought to underlie mental illness, many genes have been proposed to be implicated in the etiology of psychiatric disorders (Farrell et al., 2015). These candidate genes, implicated in processes such as neurodevelopment, dopamine signaling, or glutamate receptor signaling, as well as cell-to-cell communication and neuronal plasticity, were screened by, for example, Sanger sequencing technology in order to determine the mutational burden in cases compared to controls. The rationale that these genes and their encoded products would be involved has been historically based on their role in the biological systems hypothesized to be involved in the etiology of the disorder, such as the glutamate receptor and dopaminergic signaling pathways in schizophrenia (Abi-Dargham & Meyer, 2014) and the biological clock genes (Frank et al., 2013; Gonzalez, 2014; Karatsoreos, 2014; Landgraf, McCarthy, & Welsh, 2014) and hyperpolarization-activated cyclic nucleotide-gated channels in bipolar disorder (Kelmendi et al., 2011). What would nowadays be considered small cohorts of cases and controls were screened with mixed positive and negative results. The historical candidates for schizophrenia have recently been reviewed by expert-geneticists (Farrell et al., 2015) and, additionally, the genome-wide association (GWAS) data from the psychiatric genomics consortium (PGC) do not implicate most of the historical candidate genes as major risk factors for schizophrenia. Exceptions are several genes associated with glutamatergic signaling and the gene *DRD2* which codes for the dopamine 2 receptor, the mechanism of action of antipsychotic medication (Ripke et al., 2014). The argument for the historically difficult-to-obtain genetic signal is the bias toward the a priori biological function of the candidate genes. Consequently, there is little sound genetic evidence of involvement of these specific candidate genes in the etiology of psychiatric disorders.

Mendelian genetics

Given the clustering of psychiatric disorders in families, many family-based studies have been performed. In Mendelian genetics, linkage analysis is widely used to identify chromosomal *loci* in a pedigree which are "linked" to traits or conditions (Dawn Teare & Barrett, 2005). There were two major papers published in the late 1980s on bipolar disorder (Egeland et al., 1987) and schizophrenia (Sherrington et al., 1988) showing statistically significant linkage to chromosomes 11 and 5, respectively. After that, many researchers attempted family-based linkage analyses, which resulted in many loci found to be significantly linked with psychiatric disorders, but few of these findings were successfully replicated. A meta-analysis of family-based linkage studies in 3255 families with schizophrenia comprising 7413 cases was published describing that only suggestive linkage was found on the long arm of chromosome 5 and the long arm of chromosome 2. In Europeans specifically, the authors found suggestive linkage on the short arm of chromosome 8. The authors state however, that their analysis showed that many more loci might be linked to schizophrenia (Ng et al., 2009). With schizophrenia as the hallmark psychiatric disorder, these findings intimated the complicated genetic architecture of psychiatric disorders (Kendler, 2013).

Currently, in many fields of medicine, linkage analysis coupled with exome sequencing is applied, which allows the reading of the protein-coding fraction of the DNA, in single patients and in families with phenotypes of interest (Bonifati, 2014; Ott, Wang, & Leal, 2015). This is done in search of segregating Mendelian variants that confer a strongly increased risk of illness. The search for Mendelian variants underlying human disease continues with great success. Increasingly, more variants are being identified underlying phenotypes which were previously difficult to diagnose (Chong et al., 2015). The NIH-funded initiative *Centers for Mendelian Genomics (CMG)* has identified 956 genes of which 375 were not yet associated with human health and disease in the past 5 years. These data demonstrate Mendelian genetics is a vivid field of study. Additionally, high-penetrance coding variants provide us with the unique opportunity to study its functional consequence in a controlled laboratory environment. For variants which are in the coding regions of a gene, the functional consequences are relatively well understood and can be modeled in a laboratory setting in cell culture or in model organisms. Given the complex genetic architecture underlying psychiatric disorders indicating a polygenic etiology, it appears a common major genetic vulnerability factor for psychiatric disorders does not exist. The new generation of Mendelian genetics whereby classical family linkage studies are combined with the next-generation sequencing (NGS) technologies has the potential to unravel the genetic architecture in a family-per-family fashion. In the era of GWAS studies in psychiatry, the field of Mendelian genetics studying families with psychiatric phenotypes has been neglected. We recently published our findings identifying two independent rare Mendelian variants in the gene *CSPG4* segregating with schizophrenia using NGS coupled to linkage analysis. We found functional evidence of the pathogenicity of these variants in both fMRI imaging experiments and in vitro cell culture experiments with patient-derived oligodendrocyte experiments (de Vrij et al., 2018). From a relative-risk point of view, as well as regarding the possibilities of studying the aberrant biology in a laboratory situation, Mendelian forms of psychiatric disorders provide an excellent opportunity to gain insight into the disease pathophysiology because of the rare variants with a large effect size. As is the case for the major neurological disorders Alzheimer disease, Parkinson disease, and fronto-temporal dementia, our current medical-biological knowledge with respect to the pathophysiology of these disorders is derived from families with Mendelian pathogenic mutations.

Genome-wide association studies

The development of massive parallel genotyping assays in the form of microarray technology allowing for simultaneous inquiry of hundreds of thousands of single nucleotide polymorphisms (SNPs) led to large collaborative consortia studying complex traits and disease phenotypes such as psychiatric disorders. This methodology allows for statistical association studies of genetic markers (SNPs), which are common in the population, with membership of the patient group versus the control group, or alternatively they can be associated with a continuous trait such as height. Given the vast number of statistical tests performed, multiple testing correction has to be applied, which leads to very stringent criteria of statistical significance. The most recent iterations of GWAS for the different psychiatric disorders are led by the PGC. Regarding schizophrenia, a GWAS was performed with over 30,000 patients and more than 120,000 controls. This resulted in 108 significantly associated independent loci containing variants with relatively low effect sizes (Ripke et al., 2014). Recently, a strong association signal was found on chromosome 6 by GWAS studies for schizophrenia, partially due to variations of the complement component 4 (C4) which were associated with brain expression levels of C4 and complement component 5 (C5). The schizophrenia-associated variation was associated with higher C4 expression levels. It has been shown that C4 has a role in synaptic pruning and therefore could explain part of the pathology in schizophrenia (Sekar et al., 2016). In addition, 30 novel susceptibility loci were identified by a large study in the Chinese population. The novel loci explain 1.34% of the variance (Li et al., 2017). For bipolar disorder, a GWAS has been performed with $\sim$12,000 patients and $\sim$52,000 controls. In this study, significant association was found for markers close to *CACNA1C and ODZ4* (Psychiatric GWAS Consortium Bipolar Disorder Working Group, 2011). For major depressive disorder (MDD), no significant SNPs were identified in the 2013 GWAS comprising $\sim$9200 cases and $\sim$9500 controls in the discovery phase and $\sim$6800 cases and $\sim$50,000 controls in the replication phase (Ripke et al., 2013). However, recently 44 genetic loci have been published to be associated with the risk of MDD in a study that looked at $\sim$135,000 patients with self-reported MDD and $\sim$345,000 controls (Wray et al., 2018). For autism, it is well known that many rare monogenic syndromes exist that overlap with autism spectrum disorders (such as fragile X syndrome and tuberous sclerosis complex). Recent genetic studies investigating idiopathic autism spectrum disorder have been performed by the Simons Foundation and these studies have focused on identifying de novo mutations and are thus not present in the parents of the proband. A large recent study involving 2500 families found that if de novo mutations are combined with copy number variants (CNVs), de novo coding mutations explain 30% of the variance in the simplex cases (Iossifov et al., 2014).

Next-generation sequencing

The recently developed technologies of NGS, whereby the protein-coding fraction (whole exome sequencing, WES) or the entire genome (whole genome sequencing, WGS) can be sequenced, pose new possibilities. Recently, a large case–control association study was performed using exome sequencing data instead of SNP array data. This study was performed on $\sim$2500 cases and $\sim$2500 controls from Sweden. The authors of the study however, did not identify variants that met genome-wide significance after multiple testing correction (Purcell et al., 2014). Analysis of exomes in $>$4000 patients with schizophrenia, $<$9000 controls, and $>$1000 trios revealed significant association of loss-of-function variants in the gene *SETD1A*. In a general population cohort of $>$45,000 people, only two heterozygous loss-of-function variants were identified, suggesting strong negative selection. Identified carriers of *SETD1A* mutations were found to be associated with learning difficulties and neuropsychiatric phenotypes, leading to accumulating evidence that epigenetic dysregulation (SETD1A is a histone-lysine *N*-methyltransferase) might play a role in the pathogenesis of schizophrenia (Singh et al., 2016).

Polygenic inheritance

Historically, human molecular genetics has focused on traits and disorders with an apparent Mendelian pattern of inheritance. To date however, the field of Mendelian genetics cannot provide a molecular mechanism for most complex traits and disorders which are common in the population (Sawcer, Ban, Wason, & Dudbridge, 2010). It has been proposed that for disorders such as schizophrenia, a polygenic risk score, whereby multiple to many genetic risk alleles are taken into account, would better describe the underlying genetic architecture (Ripke et al., 2014; Szatkiewicz et al., 2014; Wray, Goddard, & Visscher, 2007). In a study with Irish families with a high incidence of schizophrenia, it was observed that the polygenic risk score was significantly elevated in both affected as well as unaffected family members (Bigdeli et al., 2014). The PGC recently published a large GWAS identifying 108 significantly associated loci with schizophrenia. Polygenic risk score profiling using these loci explained 3.4% of the variance. Depending on the dataset used (population-based, hospitalization-based, or ascertained for genetic studies), the decile with the highest burden of risk alleles has an odds ratio for affected status of 7.8, 15.0, and 20.3, respectively (Ripke et al., 2014). Currently, the polygenic risk scores for individual patients, however, are not considered suitable for use in clinical and diagnostic settings because of a lack of predictive power for an individual patient (Dudbridge, 2013; Kong et al., 2014).

Copy number variants

CNVs are deletions or duplications, but can also be insertions or translocations, of large physical segments of DNA (Iafrate et al., 2004; Sebat et al., 2004). Consequently, their presence can significantly alter gene expression and therefore has the potential to be pathogenic. Nevertheless, CNVs frequently occur in the population with an average of 1.44 CNVs per person, with a tendency for smaller CNVs to be more common. The average size in the population is ~ 205 kb (Chen et al., 2011). CNVs have been associated with psychiatric disorders, most prominently with schizophrenia (Marshall et al., 2016) and ASD (Pinto et al., 2010; Sanders et al., 2011). Studying the function of the genes comprised by the CNVs has the potential to further elucidate the etiology of psychiatric disorders.

A comparison to neurodegenerative disorders

In the field of neurodegenerative disorders, where the understanding of the disease genetics is more advanced than in psychiatry, it is widely acknowledged there are rare, usually monogenic, forms of disease as well as idiopathic forms. Although the vast majority of Alzheimer and Parkinson disease cases are considered "sporadic" with significant non-Mendelian genetic contributions, roughly 5%–15% of cases are familial and in a smaller fraction the causative genetic mutations can be identified. Most of our current understanding about the pathological mechanisms underlying these diseases is driven by models based on these rare, familial forms. Remarkably, in most cases, the clinical phenotype does not distinguish rare monogenic forms from sporadic forms (Bonifati, 2014; Klein & Westenberger, 2012; Lesage & Brice, 2009; Loy, Schofield, Turner, & Kwok, 2014; Ott et al., 2015; Papapetropoulos, Adi, Ellul, Argyriou, & Chroni, 2007; Ryan & Rossor, 2010). Since the nature of the neurodegenerative disorders is similar to psychiatric disorders: involving the brain, of pluriform etiology, and the concept that the diagnosis is most often made on the basis of clinical evaluation, we attempted to implement a methodology often used in neurology, consisting of a family-based genetic analysis involving parametric linkage and NGS followed by either the identification of other families with segregating variants in the candidate gene and/or determining a statistical enrichment of variants in the candidate gene in cases versus controls (Bonifati, 2014). Recently, this type of linkage-based NGS has been proposed as the method of choice because of its power in identifying pathogenic mutations or risk factors in families (Ott et al., 2015).

CURRENT STATE-OF-THE-ART IN THE CLINICAL PSYCHIATRIC GENETICS

With regard to clinical psychiatric genetics, the applications of WES and WGS have been limited. For neurodevelopmental disorders, most experience has been accumulated. This is likely due to major efforts to identify the etiology of early-onset neurodevelopmental disorder as soon as being possible in an attempt to cure the patient.

The impact of psychiatric disorders

Psychiatric disorders impose a strong burden on the patient, as well as on their families and significant others. Often, the psychiatric disorders are perceived as disorders of identity, whereby there appears to be a merger between the identity of the person and having a psychiatric disorder. In language, this translates to the terms "schizophrenic" or "autistic", which can be perceived as either neutral or pejorative, but nevertheless carry within them the full merger between the person and their disease. Seeman investigated the relationship between the core aspects of identity and its operationalization (name, body, religion, dress, food, gender, ethnicity) in relationship with schizophrenia and found that patients who use mental health services are more likely to make sometimes drastic changes in these areas of life (Seeman, 2017).

Genetic counseling for psychiatric disorders

The vast majority of patients with psychiatric disorders are never recommended for genetic counseling, even in the case of a strong family history. The impact a high familial load of psychiatric disorder can be significant, also beyond etiological risk (Bouwkamp, Lambregtse-Van den Berg, Kievit, & Kushner, 2017).

A genetic consultation requested by the mental healthcare team to provide expert advice regarding the genetic context of the patient and his or her family is distinct from incorporating concepts of genetic counseling into long-term treatment planning and psychotherapy administrated entirely by the mental healthcare team. Genetic counseling could be considered "a time limited, highly circumscribed form of psychotherapy" (Austin, Semaka, & Hadjipavlou, 2014). Often, feelings of guilt and shame are present in cases of illness, not least in instances of the parents of an affected (adult) child. Seymour Kessler, who wrote a lot about the psychological dynamics of genetic counseling, describes three methods to relieve stress due to feelings of guilt or shame: confession, contrition, and penance (Kessler, Kessler, & Ward, 1984). Hereby, the idea is that verbalizing the thought of guilt regardless of its degree of true relationship with the illness is beneficial.

Having a child with an early-onset (genetic) abnormality can be perceived as a narcissistic injury. Kessler writes it is an often-seen mechanism of coping with the feelings of guilt by stopping reproduction altogether in order to cope with strong feelings of guilt. This is consistent with what is seen in the clinic (Kessler, Kessler, Ward, & Opitz, 1984).

Many explanatory models hold that a psychiatric disorder is not a medical illness. Here, the distinction of illness and disease can be used. One can have a disease without being ill. Eisenberg refers to the example of psychosis which has a strong genetic-physiological basis but is expressed in developmental history and interpersonal contact. Illness is not merely medical, as put forward in the medical model, but the patient can also influence its course and recovery while maintaining the recognition that the patient is sick (Eisenberg, 1977). Genetics is part of the biopsychosocial explanatory model (Engel, 1980) but a great part of the discourse during genetic counseling should consist of putting the biological vulnerabilities conferred by a tentative genetic etiology in the proper perspective. Earlier we have seen that the genetic contribution to the development of schizophrenia in monozygotic twins is (only) ~50%. This leaves open the question as to the other contributing factors in the psychological and social domain.

In this respect, genetic counseling could be seen as a vehicle for initiating discussions of explanatory models of a patient's illness and the implications within their families. Genetic counseling has been found to be very much appreciated by patients and their family members despite the uncertainties inherent to genetic counseling (Hippman et al., 2013; Lawrence & Appelbaum, 2011). In Canada, a psychiatric genetic counseling outpatient clinic has been established, in which the early results indicate beneficial effects for patients in the areas of self-efficacy and empowerment (Inglis, Koehn, McGillivray, Stewart, & Austin, 2015). In this outpatient clinic, nonsyndromic patients are seen. Family history information is collected through telephone contact before the genetic counseling session. At the consultation, the patient or families are received. Genetic counseling outcomes and self-efficacy levels are measured before and after the consultation. The counseling itself is based on a hybrid model between education and counseling. An important theme is the explanatory model the patient has for the disease.

Jenkins and colleagues studied expert opinions with regard to genetic counseling for psychiatric disorders in the United Kingdom, where genetic counseling for psychiatric disorders is not provided. They inquired several mental health professionals and potential costs and benefits were assessed. One of the relevant findings is that mental healthcare professionals seem to be of the opinion that clinical genetics is a specialty equipped better than they are to provide genetic information, and that clinical genetic consultations allow for more time to discuss the disorder and etiology and consequences for the family in more detail and especially with more time (Jenkins & Arribas-Ayllon, 2016). Our view is that genetic counseling should be considered to be incorporated in the training curriculum for licensure of mental healthcare providers, including psychiatrists and clinical psychologists, to enhance the collaborative potential following consultative genetic counseling.

Empirical recurrence risks

One of the hallmark topics of the discussion in both prenatal and postnatal genetic counseling for psychiatric disorders is the empirical recurrence risk. People with a positive family history for mental illness frequently would want to know what the risk is that their offspring will develop symptoms, or children from affected parents would want to know what the chance is that they themselves develop symptoms.

One can distinguish recurrence risk based on the absence of prior genetic information from cases where there is genetic information available. For schizophrenia, the empirical recurrence rate sharply decreases with increased genetic distance from the proband (Table 11.1). A recent meta-analysis of 33 studies comprising 3863 offspring of parents with severe mental illness and 3158 offspring of controls documented the offspring risk of developing schizophrenia, bipolar disorder, or MDD when one of the parents is affected (Rasic, Hajek, Alda, & Uher, 2014). Aggregated risk was approximately 30% for developing one of these three disorders, and ~50% for developing any mental disorder. Although still preliminary and ideally performed using a population-based, prospective, longitudinal design, these results indicate that in the absence of any molecular genetic information, the likelihood of transgenerational transmission of mental illness is of considerable clinical significance.

The risk propagated by carriership of a CNV is anywhere between ~2 to ~70-fold higher than in the general population (Sullivan et al., 2017). With regard to CNVs, they may arise de novo and are recurrent in the population because of vulnerable loci in the genome (Marshall et al., 2016). Larger chromosomal abnormalities might be related to psychiatric disorder but are quite rare (MacIntyre et al., 2003). The posterior attributable risk for de novo CNVs is often quite high (Gershon & Alliey-Rodriguez, 2013). The latter means that when one is affected and in diagnostic testing a de novo CNV is identified, the proportion of the risk in that person which can be attributed to the de novo CNV.

Syndromic forms of mental illness, and genetic testing and counseling

The strongest known molecular risk indicator for syndromic psychiatric illness is carriership of the 22q11.2 microdeletion (Sullivan, Daly, & O'Donovan, 2012). Carriers of this microdeletion have a 30-fold increased chance to develop schizophrenia. There is however debate as to whether or not the clinical schizophrenia that results from this syndromic context is comparable to idiopathic schizophrenia (Verhoeven & Tuinier, 2008). It is well known that primary mitochondrial (Anglin, Garside, Tarnopolsky, Mazurek, & Rosebush, 2012) and other genetic syndromes [such as Prader—Willi and fragile X syndrome (Steinhausen et al., 2002)] can present with psychiatric symptoms. Although perhaps psychiatric disorders in the context of other phenotypes might be different to idiopathic mental illness, we might be able to study these cases and derive from them a theory for neurobiological underpinnings of psychiatric symptoms in general.

We recently performed a study investigating two independent cohorts of patients with chronic psychiatric disorders such as schizophrenia and autism spectrum disorders who also had intellectual disability, congenital abnormalities, and/or dysmorphic features. We identified a pathogenic or likely pathogenic genetic variant in 24% of the patients by microarray-based CNV screening (Bouwkamp et al., 2017). Despite the relatively young age of the patients, the large majority of the parents did not come to the consultation.

It is known that CNVs can have a wide range of penetrance of expressivity. Based on our results with these cohorts we recommend that genetic testing and counseling are considered in psychiatry, especially in this patient group with multiple problems. Furthermore, we recommend that testing be considered in both the presence and absence of a family history of mental illness or aspects related to the proband's phenotype. The variants might originate de novo and therefore by relying on positive history one might miss the opportunity to identify a causative mutation.

Acceptance and commitment therapy

A possible intervention to apply both in genetic counseling as well as during psychiatric diagnostic procedures is based on acceptance and commitment therapy (ACT). ACT is a behaviorally oriented psychological intervention aimed at accepting what cannot be changed and pursuing meaningful activities in the new condition. This is an evidence-based treatment developed by Hayes (Hayes, Strosahl, & Wilson, 1999), which has been shown to be effective in the treatment of both depressive and anxiety disorders as well as physical health problems (A-Tjak et al., 2015). It has been proven to be an effective intervention for health-related anxiety (Eilenberg, Fink, Jensen, Rief, & Frostholm, 2015). The core of this form of treatment is to teach patients to accept their thoughts, feelings, and behaviors as they are and have been in the past and to notice the patient as a person does not equal his or her thoughts, emotions, and behaviors. Therefore, one can make an effort to direct their energy to personal goals and values and take action on them. This increases the psychological resilience.

Aspects of this intervention can be applied during genetic counseling in order for the patient to cope with a newly received diagnosis. Genetic counselor Stephanie Broley nicely describes the application of ACT techniques in genetic counseling of the mother of a child with 22q11.2 deletion syndrome who had a lot of worries about the yet largely unknown developmental path of this variable syndrome (Broley, 2013). In a few counseling sessions, the mother's worry was addressed. In the first session, the mother expressed her worries about the future for her daughter and the uncertainty with regard to the information about the syndrome obtained from different sources. She directed much of her energy to recent surgical interventions which were medically necessary and worried about the uncertain future.

In the second session, the counselor helped the mother structure the many worries she had by clustering them in a manageable number (will she look different and be teased?, will she be intellectually disabled?, will she be able to have friends and cope socially?). The counselor then focused on the worry of the mother, which was the manner by which she coped with the situation. She could not express her worries with her significant others for various reasons. The counselor suggested she could discuss with friends and in online chat rooms with families of children who also have 22q11.2 deletion syndrome. Worrying was actively limited in time and distraction sought at those moments where it came up. In the third session, the mother indicated she felt less isolated after joining the chat rooms and applied the stop and distract technique with partial success, whereby not succeeding was also associated with stress. Cognitive defusion was suggested, whereby she would not take the worrying thought as the one and only perspective, but as just thoughts that were coming up.

Since, currently, the major psychiatric disorders such as schizophrenia, bipolar disorder, and autism are considered incurable, but manageable with pharmacological and psychological treatments, ACT-like interventions, whether in the form of ACT itself or as part of other forms of treatment, are likely to contribute to symptom relief and increased quality of life. As is often the case for neuropsychiatric disorders with or without a causative genetic basis, patients and people in their environment attempt to exert control of the situation, which is in fact uncontrollable. This can be expressed as guilt, depression, anxiety, and obsessive-compulsive symptoms. Objectively, it can be stated that neuropsychiatric disorders often come with uncertainty, regardless of the presence of a molecular genetic finding.

POINTS OF DISCUSSION

Psychiatric genetic studies have been performed since the beginning of the 20th century (Propping, 2005). Much of the progress in human molecular genetics has been paced by technological developments (Durmaz et al., 2015), that together aim to identify the risk factors underlying distinct psychiatric disorders and their shared symptomatology (Ott et al., 2015).

Family genetic studies complementary to large-scale case/control studies

The study of multiplex families has demonstrated merit in elucidating genetic determinants of human disease. By focusing on coding variants in a family-based design, it is possible to identify highly penetrant rare segregating coding variants which underlie psychopathological risk. The success of the Mendelian Genetics Consortium shows that linkage analysis coupled with exome sequencing is a valid approach for the identification of medically relevant variants in families (Chong et al., 2015). Such variants can instantaneously be used in a diagnostic genetic laboratory in diagnostic efforts for other patients and their families.

Understanding of disease etiology in neurology is derived from family studies—Why should psychiatric disorders be any different?

If the goal is to understand the pathophysiology of psychopathology, we might have to look at other related fields of study, such as neurology, in which the clinical target organ is the brain (La Spada & Ranum, 2010). A stubborn problem in frustrating investigation into the biology of psychiatric disorders is the nearly absent opportunity to obtain living diseased brain tissue. In contrast, cancer biology has advanced rapidly in part due to the widespread clinical availability of primary tumor tissue amenable to ex vivo scientific interrogation. Moreover, the lack of objective laboratory tests or imaging studies as inclusive psychiatric diagnostic criteria has also been a substantial limitation with regard to disease heterogeneity.

The genetic underpinnings of neurodegenerative diseases are also far from completely understood (Bettens, Sleegers, & Van Broeckhoven, 2010; Bonifati, 2014; Loy et al., 2014). In many respects, neurodegenerative diseases are comparable to psychiatric disorders, although there are also distinctions: they both involve the brain, are most often only clinically diagnosed, and the availability of brain imaging confirmation depends on treatment center and country of treatment. The known genes which underlie the neuropathology in the major neurodegenerative disorders were identified through family-based studies (Levy-Lahad et al., 1995; Polymeropoulos et al., 1997; Sherrington et al., 1995; Tanzi et al., 1992). After that, unrelated patients were also found to have causative mutations in these genes and even environmental influences could alter the expression levels of these genes, leading to an increased risk of developing clinical symptoms (Lesage & Brice, 2009).

More families need to be examined to sketch the landscape of mutations that may result in psychiatric disorders

The question might then arise of how many families would need to be investigated in order to identify the array of possibilities regarding genetic risk of psychiatric disorders. The theoretical number of possible coding sequence alterations which might lead to pathological brain functioning is very large, and it will probably turn out that the real number of possibilities will be very large as well. The answer remains elusive until the experiments are actually performed. Statistical geneticist Kenneth Kendler indicates in his review in Molecular Psychiatry a few years ago, that there must be some logical organization. It might be complex, but the scenarios that there are either only one or two genetic forms of a certain diagnostic entity, or that every patient has a completely independent genetic vulnerability are unlikely (Kendler, 2013). It is more likely that there is an unknown, but finite, number of genetic variations that underlies clinical psychiatric phenotypes. In other disorders where the number of loci is still increasing, there appears to be convergence on a limited number of pathways, such as is currently the case for hereditary spastic paraplegia, where there are >70 loci known which converge on <10 pathways (Salinas, Proukakis, Crosby, & Warner, 2008). For psychiatric disorders it might well be the same (Kendler, 2013). In 2017, a perspective was published with an interesting and substantiated vision, which is held intuitively by many geneticists, namely that when it comes to genetic risk for a trait or disease, the entire context of the genome matters as well as the context of the etiological cell types (Boyle, Li, & Pritchard, 2017). One could think of it as a vector sum consisting of risk and protective alleles. The composition determines the risk one has to develop a given disease, even if one does not actually have the disease. It is likely that even in the case of rare highly pathogenic mutations in families, there is still a strong influence from the context of the rest of the genome. It could be however, that a pathogenic allele might be so potent that it can lead to development of disease by itself, while in some conditions when the variants confers a weaker risk, whether or not the disease threshold is reached is influenced also by other variants and by environmental conditions. Others might have a fortunate genetic constitution and be relatively protected from developing disease. In the latter case there would be mainly alleles that have a negative risk of developing disease. In addition to this idea, also the environmental factors cannot be negated. In all situations, they play a major role (Polderman et al., 2015). In their paper describing the omnigenetic model, the authors describe that most identified genome-wide significant loci have small effect sizes and that they together explain only a small portion of the etiology. Data from the GIANT study on height demonstrates that 62% of all common genetic variants contribute to height. The median genetic variant contributes 0.14 mm, while the strongest variants contribute 1.43 mm.

The genetic architecture of childhood-onset versus adult-onset psychiatric disorders

It is likely that psychiatric disease risk is inherently polygenic, meaning that the absolute disease risk is a function of the sum of both risk and protective alleles (Wray et al., 2014). It might be that the more deleterious the types of mutations are, the earlier the onset is, and the more severe it's course. This is the case with childhood-onset schizophrenia, where the age of onset is before the age of 13 (Addington & Rapoport, 2009). More generally, this is what has been found in other brain disorders such as Parkinson's disease, with a higher likelihood to have an identifiable genetic origin with earlier onset compared to typical late-onset disease (Bonifati, 2014; Gasser, 2009). It seems that the more severe the disorder, and the earlier the age of onset, the higher the likelihood that a highly penetrant mutation lies at the foundation of the phenotype. In detailed studies of mouse models of cone-rod-homeobox protein CRX-associated retinopathies this has been confirmed by determining the gene expression levels of *Crx* by RNA sequencing and correlating them with the phenotypic severity. There was a significant association between a reduction in the level of gene expression and the severity of the phenotype. Also, it was found there was a threshold above which no apparent phenotype was observed (Ruzycki, Tran, Kefalov, Kolesnikov, & Chen, 2015). Additionally, there is a significant shared polygenic contribution to the major psychiatric classifications of ASD, schizophrenia, bipolar disorder, MDD, and alcoholism (Gandal et al., 2018). This recent study found that in terms of genetic severity ASD > SZ = BP > MDD.

Genetic testing in the psychiatric clinic in a subgroup of complex patients

A causative or likely causative rare genomic rearrangement can be diagnosed in 24% of syndromic psychiatric patients (Bouwkamp et al., 2017). Increasingly, NGS techniques are finding their way into clinical practice. The current diagnostic yield in nonsyndromic patients is too low to justify genetic testing in all nonsyndromic psychiatric patients, although other opinions exist (Rees et al., 2014). However, given the results from our study (Bouwkamp et al., 2017), we advise to start offering standardized genetic counseling and testing to all syndromic psychiatric patients and their families. The opportunity for genetic counseling and testing is readily available for patients with neurodevelopmental phenotypes and their families. The diagnostic yield is high in these cases. Exome sequencing, for example in intellectual developmental disorder, yielded a diagnosis in 28 of 41 probands, and in 18 (44%) a treatment was started specifically aimed at the identified dysfunction (Tarailo-Graovac et al., 2016).

Clinicians working in psychiatry should be trained in human molecular genetics

Genetic studies have always been performed in psychiatry, but always by experts and usually in the context of research. Initially, genetic research consisted primarily of statistical genetics in the form of epidemiology and twin studies. Figures such as concordance rates and population risk were calculated in the early 20th century (Polderman et al., 2015; Sullivan et al., 2012). Dedicated researchers also applied rapidly the emerging molecular genetic techniques to the field of psychiatry. However, in the absence of clear-cut genetic findings, knowledge about the methodology of genetics never became common knowledge amongst clinicians in the field. Human molecular genetic techniques are now increasingly becoming mainstream techniques to diagnose human disease, indicate therapeutic opportunities, improve pharmacotherapy, and to determine the presence of hereditary risk factors. Clinicians working in psychiatry should have a better understanding of the available molecular genetic techniques and be able to assess the added value for patient care if the question arises.

Genetics might change the classification system of psychiatric disorders based on the molecular disturbance rather than the clinical phenotype per se

Historically, psychiatric disorders were classified by means of their phenomenology. The current methods of human genetics may reshuffle the system of classification of psychiatric disorders. In the recent cross-disorder genetic studies (where different diagnoses are pooled together), it has been observed that certain genetic risk factors confer risk for developing several psychiatric disorders (Smoller et al., 2013). With the collective years of experience in the field of psychiatric genetics, we have learned that genes are often not solely responsible for a given function, but instead affected an entire biological cascade of events which leads to a vulnerability at a systems level (Kendler, 2013).

CONCLUDING REMARKS

There are identifiable genetic factors that are associated with an increased likelihood of developing psychiatric disorders. In concert with other risk factors in the environment, they can result in manifestation of the phenotype. The work so far indicates that we are still in the early stages of discovering the genetic architecture underlying psychiatric disorders. A lot of work remains ahead to identify other major genetic risk factors that predispose to psychiatric disorders, to model how the genetic architecture underlies clinical phenotypes, and how we can use this newly acquired information in the discovery of therapeutic agents and apply these to patient care and cure.

REFERENCES

Abi-Dargham, A., & Meyer, J. M. (2014). Schizophrenia: The role of dopamine and glutamate. *The Journal of Clinical Psychiatry, 75*, 274–275.

Addington, A. M., & Rapoport, J. L. (2009). The genetics of childhood-onset schizophrenia: When madness strikes the prepubescent. *Current Psychiatry Reports, 11*, 156–161.

American Psychiatric Association. (2013). *Diagnostic and statistical manual of mental disorders: DSM-5* (5th ed.). Washington, DC: APA.

Anglin, R. E., Garside, S. L., Tarnopolsky, M. A., Mazurek, M. F., & Rosebush, P. I. (2012). The psychiatric manifestations of mitochondrial disorders: A case and review of the literature. *The Journal of Clinical Psychiatry, 73*, 506–512.

A-Tjak, J. G. L., Davis, M. L., Morina, N., Powers, M. B., Smits, J. A. J., & Emmelkamp, P. M. G. (2015). A meta-analysis of the efficacy of acceptance and commitment therapy for clinically relevant mental and physical health problems. *Psychotherapy and Psychosomatics, 84*, 30–36.

Austin, J., Semaka, A., & Hadjipavlou, G. (2014). Conceptualizing genetic counseling as psychotherapy in the era of genomic medicine. *Journal of Genetic Counseling, 23*, 903–909.

Austin, J. C., & Peay, H. L. (2006). Applications and limitations of empiric data in provision of recurrence risks for schizophrenia: A practical review for healthcare professionals providing clinical psychiatric genetics consultations. *Clinical Genetics, 70*, 177–187.

Bettens, K., Sleegers, K., & Van Broeckhoven, C. (2010). Current status on Alzheimer disease molecular genetics: From past, to present, to future. *Human Molecular Genetics, 19*, R4–R11.

Bigdeli, T. B., Bacanu, S.-A., Webb, B. T., Walsh, D., O'Neill, F. A., Fanous, A. H., Riley, B. P., ... Kendler, K. S. (2014). Molecular validation of the schizophrenia spectrum. *Schizophrenia Bulletin, 40*, 60–65.

Bleuler, E. (1950). *Dementia praecox or the group of schizophrenias*. Oxford: International Universities Press.

Bonifati, V. (2014). Genetics of Parkinson's disease—State of the art, 2013. *Parkinsonism and Related Disorders, 20*(Suppl. 1), S23–S28.

Bouwkamp, C. G., Kievit, A. J. A., Markx, S., Friedman, J. I., van Zutven, L., van Minkelen, R., Vrijenhoek, T., ... Kushner, S. A. (2017). Copy number variation in syndromic forms of psychiatric illness: The emerging value of clinical genetic testing in psychiatry. *The American Journal of Psychiatry, 174*, 1036–1050.

Bouwkamp, C. G., Kievit, A. J. A., Olgiati, S., Breedveld, G. J., Coesmans, M., Bonifati, V., & Kushner, S. A. (2017). A balanced translocation disrupting BCL2L10 and PNLDC1 segregates with affective psychosis. *American Journal of Medical Genetics Part B: Neuropsychiatric Genetics, 174*, 214–219.

Bouwkamp, C. G., Lambregtse-Van den Berg, M. P., Kievit, A. J. A., & Kushner, S. A. (2017). Psychodynamic consequences of a family history with psychiatric disorders. *Tijdschrift Voor Psychiatrie, 59*, 474–481.

Boyle, E. A., Li, Y. I., & Pritchard, J. K. (2017). An expanded view of complex traits: From polygenic to omnigenic. *Cell, 169*, 1177–1186.

Broley, S. (2013). Acceptance and commitment therapy in genetic counselling: A case study of recurrent worry. *Journal of Genetic Counseling, 22*, 296–302.

Chen, W., Hayward, C., Wright, A. F., Hicks, A. A., Vitart, V., Knott, S., Wild, S. H., ... Porteous, D. V. (2011). Copy number variation across European populations. *PLoS One, 6*, e23087.

Chong, J. X., Buckingham, K. J., Jhangiani, S. N., Boehm, C., Sobreira, N., Smith, J. D., Harrell, T. M., ... Bamshad, M. J. (2015). The genetic basis of Mendelian phenotypes: Discoveries, challenges, and opportunities. *American Journal of Human Genetics, 97*, 199–215.

Dawn Teare, M., & Barrett, J. H. (2005). Genetic linkage studies. *Lancet, 366,* 1036–1044.

de Vrij, F. M., Bouwkamp, C. G., Gunhanlar, N., Shpak, G., Lendemeijer, B., Baghdadi, M., Gopalakrishna, S., ... Kushner, S. A. (2018). Candidate CSPG4 mutations and induced pluripotent stem cell modeling implicate oligodendrocyte progenitor cell dysfunction in familial schizophrenia. *Molecular Psychiatry.* Available from https://doi.org/10.1038/s41380-017-0004-2.

Dudbridge, F. (2013). Power and predictive accuracy of polygenic risk scores. *PLoS Genetics, 9,* e1003348.

Durmaz, A. A., Karaca, E., Demkow, U., Toruner, G., Schoumans, J., & Cogulu, O. (2015). Evolution of genetic techniques: Past, present, and beyond. *BioMed Research International, 2015.* Available from https://doi.org/10.1155/2015/461524.

Egeland, J. A., Gerhard, D. S., Pauls, D. L., Sussex, J. N., Kidd, K. K., Allen, C. R., Hostetter, A. M., & Housman, D. E. (1987). Bipolar affective disorders linked to DNA markers on chromosome 11. *Nature, 325,* 783–787.

Eilenberg, T., Fink, P., Jensen, J. S., Rief, W., & Frostholm, L. (2015). Acceptance and commitment group therapy (ACT-G) for health anxiety: A randomized controlled trial. *Psychological Medicine, 2012,* 1–13.

Eisenberg, L. (1977). Disease and illness: Distinctions between professional and popular ideas of sickness. *Culture, Medicine and Psychiatry, 1,* 9–23.

Engel, G. L. (1980). The clinical application of the biopsychosocial model. *The American Journal of Psychiatry, 137,* 535–544.

Farrell, M. S., Werge, T., Sklar, P., Owen, M. J., Ophoff, R. A., O'Donovan, M. C., Corvin, A., ... Sullivan, P. F. (2015). Evaluating historical candidate genes for schizophrenia. *Molecular Psychiatry, 20,* 555–562.

Frank, E., Sidor, M. M., Gamble, K. L., Cirelli, C., Sharkey, K. M., Hoyle, N., Tikotzky, L., ... Hasler, B. P. (2013). Circadian clocks, brain function, and development. *Annals of the New York Academy of Sciences, 1306,* 43–67.

Gandal, M. J., Haney, J. R., Parikshak, N. N., Leppa, V., Ramaswami, G., Hartl, C., Schork, A. J., ... Geschwind, D. H. (2018). Shared molecular neuropathology across major psychiatric disorders parallels polygenic overlap. *Science, 359,* 693–697.

Gasser, T. (2009). Mendelian forms of Parkinson's disease. *Biochimica et Biophysica Acta, 1792,* 587–596.

Gershon, E. S., & Alliey-Rodriguez, N. (2013). New ethical issues for genetic counseling in common mental disorders. *The American Journal of Psychiatry, 170,* 968–976.

Gonzalez, R. (2014). The relationship between bipolar disorder and biological rhythms. *The Journal of Clinical Psychiatry, 75,* e323–e331.

Hayes, S. C., Strosahl, K. D., & Wilson, K. G. (1999). *Acceptance and commitment therapy: An experiential approach to behavior change.* New York, NY: Guilford Press.

Hippman, C., Lohn, Z., Ringrose, A., Inglis, A., Cheek, J., & Austin, J. C. (2013). "Nothing is absolute in life": Understanding uncertainty in the context of psychiatric genetic counseling from the perspective of those with serious mental illness. *Journal of Genetic Counseling, 22,* 625–632.

Iafrate, A. J., Feuk, L., Rivera, M. N., Listewnik, M. L., Donahoe, P. K., Qi, Y., Scherer, S. W., & Lee, C. (2004). Detection of large-scale variation in the human genome. *Nature Genetics, 36,* 949–951.

Inglis, A., Koehn, D., McGillivray, B., Stewart, S. E., & Austin, J. (2015). Evaluating a unique, specialist psychiatric genetic counseling clinic: Uptake and impact. *Clinical Genetics, 87,* 218–224.

Iossifov, I., O'Roak, B. J., Sanders, S. J., Ronemus, M., Krumm, N., Levy, D., Stessman, H. A., ... Wigler, M. (2014). The contribution of de novo coding mutations to autism spectrum disorder. *Nature, 515,* 216–221.

Jenkins, S., & Arribas-Ayllon, M. (2016). Genetic counselling for psychiatric disorders: Accounts of psychiatric health professionals in the United Kingdom. *Journal of Genetic Counseling, 25,* 1243–1255.

Karatsoreos, I. N. (2014). Links between circadian rhythms and psychiatric disease. *Frontiers in Behavioral Neuroscience, 8,* 162.

Kelmendi, B., Holsbach-Beltrame, M., McIntosh, A. M., Hilt, L., George, E. D., Kitchen, R. R., Carlyle, B. C., ... Simen, A. A. (2011). Association of polymorphisms in HCN4 with mood disorders and obsessive compulsive disorder. *Neuroscience Letters, 496,* 195–199.

Kendler, K. S. (2013). What psychiatric genetics has taught us about the nature of psychiatric illness and what is left to learn. *Molecular Psychiatry, 18,* 1058–1066.

Kendler, K. S., McGuire, M., Gruenberg, A. M., & Walsh, D. (1994). Clinical heterogeneity in schizophrenia and the pattern of psychopathology in relatives: Results from an epidemiologically based family study. *Acta Psychiatrica Scandinavica, 89,* 294–300.

Kessler, S., Kessler, H., & Ward, P. (1984). Psychological aspects of genetic counseling. III. Management of guilt and shame. *American Journal of Medical Genetics, 17*, 673—697.

Kessler, S., Kessler, H., Ward, P., & Opitz, J. M. (1984). Psychological aspects of genetic counseling. III. Management of guilt and shame. *American Journal of Medical Genetics, 17*, 673—697.

Klein, C., & Westenberger, A. (2012). Genetics of Parkinson's disease. *Cold Spring Harbor Perspectives in Medicine, 2*, a008888.

Kong, S. W., Lee, I.-H., Leshchiner, I., Krier, J., Kraft, P., Rehm, H. L., Green, R. C., . . . MacRae, C. A. (2014). Summarizing polygenic risks for complex diseases in a clinical whole-genome report. *Genetics in Medicine, 17*. Available from https://doi.org/10.1038/gim.2014.143.

La Spada, A., & Ranum, L. P. W. (2010). Molecular genetic advances in neurological disease: Special review issue. *Human Molecular Genetics, 19*, R1—R3.

Landgraf, D., McCarthy, M. J., & Welsh, D. K. (2014). Circadian clock and stress interactions in the molecular biology of psychiatric disorders. *Current Psychiatry Reports, 16*, 483.

Lawrence, R. E., & Appelbaum, P. S. (2011). Genetic testing in psychiatry: A review of attitudes and beliefs. *Psychiatry, 74*, 315—331.

Lesage, S., & Brice, A. (2009). Parkinson's disease: From monogenic forms to genetic susceptibility factors. *Human Molecular Genetics, 18*, R48—R59.

Levy-Lahad, E., Wasco, W., Poorkaj, P., Romano, D. M., Oshima, J., Pettingell, W. H., Yu, C. E., . . . (1995). Candidate gene for the chromosome 1 familial Alzheimer's disease locus. *Science, 269*, 973—977.

Li, Z., Chen, J., Yu, H., He, L., Xu, Y., Zhang, D., Yi, Q., . . . Shi, Y. (2017). Genome-wide association analysis identifies 30 new susceptibility loci for schizophrenia. *Nature Genetics, 49*, 1576—1583.

Loy, C. T., Schofield, P. R., Turner, A. M., & Kwok, J. B. J. (2014). Genetics of dementia. *Lancet (London, England), 383*, 828—840.

MacIntyre, D. J., Blackwood, D. H. R., Porteous, D. J., Pickard, B. S., & Muir, W. J. (2003). Chromosomal abnormalities and mental illness. *Molecular Psychiatry, 8*, 275—287.

Marneros, A. (2008). Psychiatry's 200th birthday. *British Journal of Psychiatry, 193*, 1—3.

Marshall, C. R., Howrigan, D. P., Merico, D., Thiruvahindrapuram, B., Wu, W., Greer, D. S., Antaki, D., . . . Sebat, J. (2016). Contribution of copy number variants to schizophrenia from a genome-wide study of 41,321 subjects. *Nature Genetics, 49*, 27—35.

Nasser, M. (1987). Psychiatry in Ancient Egypt. *Psychiatric Bulletin, 11*, 420—422.

Ng, M. Y. M., Levinson, D. F., Faraone, S. V., Suarez, B. K., DeLisi, L. E., Arinami, T., Riley, B., . . . Lewis, C. M. (2009). Meta-analysis of 32 genome-wide linkage studies of schizophrenia. *Molecular Psychiatry, 14*, 774—785.

Ott, J., Wang, J., & Leal, S. M. (2015). Genetic linkage analysis in the age of whole-genome sequencing. *Nature Reviews Genetics, 16*, 275—284.

Papapetropoulos, S., Adi, N., Ellul, J., Argyriou, A. A., & Chroni, E. (2007). A prospective study of familial versus sporadic Parkinson's disease. *Neurodegenerative Diseases, 4*, 424—427.

Perälä, J., Suvisaari, J., Saarni, S. I., Kuoppasalmi, K., Isometsä, E., Pirkola, S., Partonen, T., . . . Lönnqvist, J. (2007). Lifetime prevalence of psychotic and bipolar i disorders in a general population. *Archives of General Psychiatry, 64*, 19.

Perugi, G., Fornaro, M., & Akiskal, H. S. (2011). Are atypical depression, borderline personality disorder and bipolar II disorder overlapping manifestations of a common cyclothymic diathesis? *World Psychiatry, 10*, 45—51.

Pinto, D., Pagnamenta, A. T., Klei, L., Anney, R., Merico, D., Regan, R., Conroy, J., . . . Betancur, C. (2010). Functional impact of global rare copy number variation in autism spectrum disorders. *Nature, 466*, 368—372.

Polderman, T. J. C., Benyamin, B., de Leeuw, C. A., Sullivan, P. F., van Bochoven, A., Visscher, P. M., & Posthuma, D. (2015). Meta-analysis of the heritability of human traits based on fifty years of twin studies. *Nature Genetics, 47*, 702—709.

Polymeropoulos, M. H., Lavedan, C., Leroy, E., Ide, S. E., Dehejia, A., Dutra, A., Pike, B., . . . Nussbaum, R. L. (1997). Mutation in the alpha-synuclein gene identified in families with Parkinson's disease. *Science, 276*, 2045—2047.

Propping, P. (2005). The biography of psychiatric genetics: From early achievements to historical burden, from an anxious society to critical geneticists. *American Journal of Medical Genetics B: Neuropsychiatric Genetics, 136B*, 2—7.

Psychiatric GWAS Consortium Bipolar Disorder Working Group. (2011). Large-scale genome-wide association analysis of bipolar disorder identifies a new susceptibility locus near ODZ4. *Nature Genetics, 43*, 977–983.

Purcell, S. M., Moran, J. L., Fromer, M., Ruderfer, D., Solovieff, N., Roussos, P., O'Dushlaine, C., . . . Sklar, P. (2014). A polygenic burden of rare disruptive mutations in schizophrenia. *Nature, 506*, 185–190.

Rasic, D., Hajek, T., Alda, M., & Uher, R. (2014). Risk of mental illness in offspring of parents with schizophrenia, bipolar disorder, and major depressive disorder: A meta-analysis of family high-risk studies. *Schizophrenia Bulletin, 40*, 28–38.

Rees, E., Walters, J. T. R., Georgieva, L., Isles, A. R., Chambert, K. D., Richards, A. L., Mahoney-Davies, G., . . . Kirov, G. (2014). Analysis of copy number variations at 15 schizophrenia-associated loci. *British Journal of Psychiatry, 204*, 108–114.

Rice, C. E., Rosanoff, M., Dawson, G., Durkin, M. S., Croen, L. A., Singer, A., & Yeargin-Allsopp, M. (2012). Evaluating changes in the prevalence of the autism spectrum disorders (ASDs). *Public Health Reviews, 34*, 1–22.

Ripke, S., Neale, B. M., Corvin, A., Walters, J. T. R., Farh, K.-H., Holmans, P. A., Lee, P., . . . O'Donovan, M. C. (2014). Biological insights from 108 schizophrenia-associated genetic loci. *Nature, 511*, 421–427.

Ripke, S., Wray, N. R., Lewis, C. M., Hamilton, S. P., Weissman, M. M., Breen, G., 'Byrne, E. M., . . . Sullivan, P. F. (2013). A mega-analysis of genome-wide association studies for major depressive disorder. *Molecular Psychiatry, 18*, 497–511.

Ruzycki, P. A., Tran, N. M., Kefalov, V. J., Kolesnikov, A. V., & Chen, S. (2015). Graded gene expression changes determine phenotype severity in mouse models of CRX-associated retinopathies. *Genome Biology, 16*, 171.

Ryan, N. S., & Rossor, M. N. (2010). Correlating familial Alzheimer's disease gene mutations with clinical phenotype. *Biomarkers in Medicine, 4*, 99–112.

Salinas, S., Proukakis, C., Crosby, A., & Warner, T. T. (2008). Hereditary spastic paraplegia: Clinical features and pathogenetic mechanisms. *Lancet Neurology, 7*, 1127–1138.

Sanders, S. J., Ercan-Sencicek, A. G., Hus, V., Luo, R., Murtha, M. T., Moreno-De-Luca, D., Chu, S. H., . . . State, M. W. (2011). Multiple recurrent de novo CNVs, including duplications of the 7q11.23 Williams syndrome region, are strongly associated with autism. *Neuron, 70*, 863–885.

Sawcer, S., Ban, M., Wason, J., & Dudbridge, F. (2010). What role for genetics in the prediction of multiple sclerosis? *Annals of Neurology, 67*, 3–10.

Schulze, T. G., Fangerau, H., & Propping, P. (2004). From degeneration to genetic susceptibility, from eugenics to genethics, from Bezugsziffer to LOD score: The history of psychiatric genetics. *International Review of Psychiatry, 16*, 246–259.

Sebat, J., Lakshmi, B., Troge, J., Alexander, J., Young, J., Lundin, P., Månér, S., . . . Wigler, M. (2004). Large-scale copy number polymorphism in the human genome. *Science, 305*, 525–528.

Seeman, M. V. (2017). Identity and schizophrenia: Who do I want to be? *World Journal of Psychiatry, 7*, 1–7.

Sekar, A., Bialas, A. R., de Rivera, H., Davis, A., Hammond, T. R., Kamitaki, N., Tooley, K., . . . McCarroll, S. A. (2016). Schizophrenia risk from complex variation of complement component 4. *Nature, 530*, 177–183.

Sherrington, R., Brynjolfsson, J., Petursson, H., Potter, M., Dudleston, K., Barraclough, B., Wasmuth, J., . . . Gurling, H. (1988). Localization of a susceptibility locus for schizophrenia on chromosome 5. *Nature, 336*, 164–167.

Sherrington, R., Rogaev, E. I., Liang, Y., Rogaeva, E. A., Levesque, G., Ikeda, M., Chi, H., . . . St George-Hyslop, P. H. (1995). Cloning of a gene bearing missense mutations in early-onset familial Alzheimer's disease. *Nature, 375*, 754–760.

Singh, T., Kurki, M. I., Curtis, D., Purcell, S. M., Crooks, L., McRae, J., Suvisaari, J., . . . Barrett, J. C. (2016). Rare loss-of-function variants in SETD1A are associated with schizophrenia and developmental disorders. *Nature Neuroscience, 19*, 571–577.

Smoller, J. W., Craddock, N., Kendler, K., Lee, P. H., Neale, B. M., Nurnberger, J. I., Ripke, S., et al. (2013). Identification of risk loci with shared effects on five major psychiatric disorders: A genome-wide analysis. *Lancet, 381*, 1371–1379.

Steinhausen, H.-C., Von Gontard, A., Spohr, H.-L., Hauffa, B. P., Eiholzer, U., Backes, M., Willms, J., & Malin, Z. (2002). Behavioral phenotypes in four mental retardation syndromes: Fetal alcohol syndrome, Prader-Willi syndrome, fragile X syndrome, and tuberosis sclerosis. *American Journal of Medical Genetics, 111*, 381–387.

Substance Abuse and Mental Health Services Administration, Results from the 2013 National Survey on Drug Use and Health: Mental Health Findings, 2014, NSDUH Series H-49, HHS Publication No. (SMA) 14-4887, Substance Abuse and Mental Health Services Administration, Rockville, MD.

Sullivan, P. F., Agrawal, A., Bulik, C. M., Andreassen, O. A., Børglum, A. D., Breen, G., Cichon, S., ... O'Donovan, M. C. (2017). Psychiatric genomics: An update and an agenda. *The American Journal of Psychiatry, 175*, 15–27, appi.ajp.2017.1.

Sullivan, P. F., Daly, M. J., & O'Donovan, M. (2012). Genetic architectures of psychiatric disorders: The emerging picture and its implications. *Nature Reviews Genetics, 13*, 537–551.

Syed, I. B. (2002). Islamic Medicine: 1000 years ahead of its times. *Journal of the Islamic Medical Association of North America, 2*, 2–9.

Szatkiewicz, J. P., O'Dushlaine, C., Chen, G., Chambert, K., Moran, J. L., Neale, B. M., Fromer, M., ... Sullivan, P. F. (2014). Copy number variation in schizophrenia in Sweden. *Molecular Psychiatry, 19*, 762–773.

Tanzi, R. E., Vaula, G., Romano, D. M., Mortilla, M., Huang, T. L., Tupler, R. G., Wasco, W., ... (1992). Assessment of amyloid beta-protein precursor gene mutations in a large set of familial and sporadic Alzheimer disease cases. *American Journal of Human Genetics, 51*, 273–282.

Tarailo-Graovac, M., Shyr, C., Ross, C. J., Horvath, G. A., Salvarinova, R., Ye, X. C., Zhang, L. H., ... Alfadhel, M. (2016). Exome sequencing and the management of neurometabolic disorders. *New England Journal of Medicine, 374*, 2246–2255.

van Os, J., & Kapur, S. (2009). Schizophrenia. *Lancet, 374*, 635–645.

Verhoeven, W. M. A., & Tuinier, S. (2008). Clinical perspectives on the genetics of schizophrenia: A bottom-up orientation. *Neurotoxicity Research, 14*, 141–150.

WHO. (1992). *The international classification of diseases: Classification of* mental and behavioural disorders: *Clinical* descriptions and diagnostic guidelines. Geneva: WHO.

Wray, N., Goddard, M., & Visscher, P. (2007). Prediction of individual genetic risk to disease from genome-wide association studies. *Genome Research, 17*, 1520–1528.

Wray, N. R., Lee, S. H., Mehta, D., Vinkhuyzen, A. A. E., Dudbridge, F., & Middeldorp, C. M. (2014). Research review: Polygenic methods and their application to psychiatric traits. *Journal of Child Psychology and Psychiatry, 55*, 1068–1087.

Wray, N. R., Ripke, S., Mattheisen, M., Trzaskowski, M., Byrne, E. M., Abdellaoui, A., Adams, M. J., ... Sullivan, P. F. (2018). Genome-wide association analyses identify 44 risk variants and refine the genetic architecture of major depression. *Nature Genetics, 50*, 668–681.

FURTHER READING

Bonke, B., Tibben, A., Lindhout, D., Clarke, A. J., & Stijnen, T. (2005). Genetic risk estimation by healthcare professionals. *The Medical Journal of Australia, 182*, 116–118.

CHAPTER 12

Opportunistic Genomic Screening: Ethical Exploration*

Guido de Wert and Wybo Dondorp
Faculty of Health, Medicine and the Life Sciences Health, Ethics and Society, Maastricht University, Maastricht, The
Netherlands

INTRODUCTION

Diagnostic testing based on next-generation sequencing (NGS) is increasingly used in patient care to clarify the cause of complex disorders. Given the falling price of sequencing technology it is expected that in the near future most patients with an indication for genetic testing will have their exomes or genomes sequenced. This has led to debate about whether the analysis of raw NGS data thus obtained should be targeted to the clinical indication at hand or widened to include mutations or risk factors for other (clinically relevant) conditions beyond that indication. Like individual experts, professional bodies have different views. The European Society of Human Genetics (ESHG) has insisted that genomic analysis should be as targeted as possible, at least for the time being (van El et al., 2013). But the American College of Medical Genetics and Genomics (ACMG) recommends to routinely analyze a series of "actionable mutations" in each case of clinical sequencing, a strategy termed "opportunistic genomic screening" (OGS) (Green et al., 2013). While this proposal has led to considerable debate in the Unites States, a truly European discussion still needs to be started. Most recently, just before the submission of the current ethical reflection, the French Society of Predictive and Personalized Medicine (SFMPP) published its guidelines on "reporting secondary findings of genome sequencing in cancer genes" (Pujol et al., 2018). The current reflection aims to stimulate a wider normative debate on OGS. Is OGS to be regarded as ethically (and legally) acceptable or even recommendable, and if so, on which conditions? We use the ACMG Recommendations as a frame of reference for further debate and ethical reflection, but will occasionally refer also to the SFMPP's proposal. This ethical reflection is to a large extent explorative, aimed at

* This chapter was written with funding from ZonMw, project nr. 80-84600-98-3002 ("ELSI Personalised Medicine"). An early draft was presented by the first author at the Farewell Symposium for Prof. Dr Aad Tibben, Leiden University Medical Center, 15 December 2017.

Clinical Genome Sequencing
DOI: https://doi.org/10.1016/B978-0-12-813335-4.00012-X

203

identifying central ethical issues for further debate and analysis, thereby contributing to moral agenda-setting. But let us first sketch the wider horizon.

THE WIDER HORIZON

The rise of personalized medicine

Modern medicine is highly dynamic, and is now in a process of developing towards what is often termed personalized medicine (PM). PM seems to be a buzzword, with a quite vague meaning. The ends and means of PM as described in the literature are rather divergent. In the context of a recent European project on PM, this was defined as follows: "PM seeks to improve tailoring and timing of preventive and therapeutic measures by utilizing biological information and biomarkers on the level of disease pathways, genetics, proteomics as well as metabolomics" (Schleidgen, Klingler, Bertram, Rogowski, & Marckmann, 2015). Clearly, PM, thus understood, covers a wide range of research, clinical, and public health activities. Quite often, however, people seem to reduce PM to genomics-based therapeutic interventions, especially in oncology. Although the authors involved in the same European project—rightly—argue in favor of a broad definition, their recommendations document seems to reduce PM largely to its therapeutic applications in disregard of its preventive potential (Schildmann et al., 2015).

Moreover, the term PM itself is contested, also from an ethical perspective. The German philosopher Vollmann, for example, considers the term PM to be "incorrect and misleading," the main aim of which is "to achieve a positive image." He wants to "debunk this questionable advertising strategy" (Vollmann, 2015). For both medical and ethical reasons, many commentators prefer terms like "tailored" or "precision" medicine. But, interestingly, others hold on to the term PM precisely on ethical grounds: "there are ethical reasons why it is potentially useful to continue using the term *personalised* medicine, as the use of this term offers the possibility not only to assume the favoured aims as set by biomedicine, but also to critically question the practice regarding its personalization, and thus its patient-centredness (...)" (Woehlke, Perry, & Schicktanz, 2015).

While acknowledging the critique, we stick to the term PM, stressing that it should not be used in a reductionist way: from a biomedical perspective, PM entails more than genomics-based therapeutic interventions, while from an ethical perspective, the person should not be reduced to the body.

NGS: a catalyst for PM

NGS allows the determination of the complete order of base pairs that make up the genome of a cell or an organism (cf. Chapter 2: Genome Sequencing and Individual Responses to Results). Sequencing an individual's complete genome is becoming cheaper each and every year. It is expected to cost less than one thousand dollars soon. If the technology becomes sufficiently cheap, NGS will likely become the standard testing approach. But sequencing is only the first step of the procedure. The second step is the analysis, in which raw sequencing data are turned into meaningful genomic information. The analysis step can either be targeted or nontargeted. For targeted analysis, filters are being used to selectively analyze parts of the genome. This helps to limit the generation of unsought genetic information. However, nontargeted analysis may help finding a diagnosis in cases where the specific molecular background is not or insufficiently clear. At present, this approach is most often based on whole exome rather than whole genome sequencing, but it is expected that this will change in the near future.

Apart from research settings, NGS-based testing can either be used for diagnostic purposes: to find a diagnosis that explains the patient's clinical features, or as a form of screening, defined as the *unsolicited* offer of a test to asymptomatic people with *no indication* for the test at hand. Such NGS-based screening would generate a "personal genomic database" "to be used to deliver" "personalized medicine" at different stages of life, in the form of, for example, personalized medication, personalized life-style advice, or timely surgery, for the purpose of either primary or secondary prevention. Likewise, NGS-based testing could be used in the context of reproductive screening. Clearly, the ethics of such screening programs is beyond the scope of this chapter. What is important for the present reflection is that the boundaries between diagnostic testing (on indication) and screening are blurring. OGS may well be a good example of such blurring.

OGS: THE ACMG'S AND SFMPP'S PROPOSALS
The concept and rationale of OGS

Forms of genetic (or genomic) screening can be categorized in different, overlapping, ways. Firstly, in terms of the distinction just mentioned, between reproductive and nonreproductive genetic screening. Secondly, in terms of the timing of screening during the human life course. Screening can be offered postnatally (to minors or adults), prenatally, or even at the preimplantation or preconception stage. And thirdly, a distinction can be made between universal screening, offered to all people or to all members of a subset of the population, for example, all pregnant women or all newborns, and selective screening, offered to a predetermined category of, for example, pregnant women. OGS is also a form of selective screening; it can be defined as the selective offer of a test to those who happen to be present or who happen to be already in a process of undergoing another test for a different reason (see below for a discussion of the inclusion criterion used).

Some may be willing to make a further distinction between two OGS-scenarios: (A) OGS as offered by professionals to the patient (or to the parents or other representatives in the case of minors or other incompetent patients) and (B) OGS as requested by patients or representatives. Type A meets both defining characteristics of screening (an unsolicited offer, to people with no indication for the screening test at hand), while type B meets only the latter criterion. We postpone a further discussion about the normative relevance of this further distinction and focus our current reflection on type A OGS (in line with the ACMG's proposal).

We focus on nonreproductive OGS, acknowledging that relevant clinical applications of OGS may be found in the reproductive context. Think of, for example, (1) testing for numerical chromosomal aberrations in the context of Preimplantation Genetic Diagnosis (PGD) for Mendelian disorders, (2) wider screening for Mendelian disorders in the context of prenatal diagnosis of, for example, Down syndrome, and, possibly, (3) the offering of preconception carrier screening (PCS) to all applicants of assisted reproduction. Obviously, the latter can only be understood as OGS if fertile couples do *not* get the offer of PCS.

Moreover, it is important to note that nonreproductive OGS may lead to information that is directly relevant for reproductive decision-making either by the testee him- or herself or his/her close relatives.

The ACMG's proposal

We briefly summarize the ACMG's original (1.0) and the ACMG's amended Recommendations (2.0):

ACMG/OGS 1.0

The ACMG 2013 proposal recommends that laboratories performing clinical exome and genome sequencing seek and report to the physician a minimum list of so-called incidental or secondary findings (IF/SF) of highly penetrant, actionable genetic mutations and variants (Green et al., 2013). The ACMG defines IF as "the results of a deliberate search for (likely) pathogenic alterations in genes that are not apparently relevant to a diagnostic indication for which the sequencing was ordered." Depending upon the specific genetic risk factor or variant, carriers may make use of individualized preventative options, including early MRI monitoring, colonoscopy, prophylactic surgery, and ICD. These IF should be sought and reported in all clinical sequencing regardless of the indication for which the clinical sequencing was ordered and irrespective of the age of the patient. The genes and variants included so far represent three basic categories: genetic risks for cancer, genetic risks for cardiovascular diseases, and (a few) variants causing adverse reactions to anesthetics. The enlisted mutations or variants are labeled as known pathogenic (KP) and expected pathogenic (EP). In cases where evidence was lacking, the ACMG Working Party (WP) drew upon the clinical judgment of its membership and some ad hoc reviewers. Although the WP recommended that only variants with a higher likelihood of causing disease be reported as IF, it recognized "that there are limited data available in many cases to make this assessment." While the minimum list, based on a consensus-driven assessment of clinical validity and utility, originally entailed 57 of genes (later revised to 56), the list has been enlarged to 63. The ACMG encourages the creation of an ongoing process of data collection and recommends refining and updating this list at least annually. Clearly, the ACMG expects that the minimum list will need regular extension.

The normative framework of the original ACMG proposal can be briefly summarized as: "the laboratory's and doctor's duty to prevent harm trumps the principle of respect for autonomy." Patients are expected to accept the test package; if they would refuse the OGS part of the test-package, they will get no testing at all—even though they have an indication.

ACMG/OGS 2.0

The original OGS proposal generated rather critical reactions, regarding both the terms in which it was framed and the normative framework behind it (see ACMG/OGS 1.0 proposal). With regard to terminology, the amended version changes IF into SF (ACMG Board of Directors, 2015). With regard to the normative framework: the amended version proposes an "opt out" system. Patients should, the ACMG now argues, be free to refuse the OGS part of the test package without spoiling access to the indicated genetic test.

The SFMPP's proposal

Stating that guidelines for the handling of SF are urgently needed, the SFMPP published its guidelines for SF regarding cancer-associated genes in adults in August 2018 (Pujol et al., 2018). The SFMPP considers its work to be a first step toward standardized guidelines in France and Europe for SFs related to cancer-predisposing genes. The Society defines SF as "the results of a deliberate or incidental screening for alterations in genes that are not relevant to the diagnostic indication for which the screening was ordered." Based on the criterion of "actionability" (available screening or prevention strategies), the risk evaluation and the level of evidence, the SFMPP recommends to deliver information on 36 (so-called "class 1") genes related to cancers in adults. There is substantial overlap between the ACMG and the SFMPP lists of actionable genes predisposing for cancer, but the SFMPP's list includes additional genes, like *PALB2*, which has recently been identified as a major predisposing gene for breast cancer. Like the ACMG, the SFMPP recommends that the list should be regularly updated according to the evolution of knowledge. Nineteen genes responsible for pediatric-onset cancer were excluded, as the topic of OGS in children will require specific considerations. Central to the SFMPP's normative framework is Recommendation no. 4: "The patient's autonomy and desire to know or to ignore such SF results must be respected. Patients could decline at any time to be informed about these SFs, even if they previously gave their approval." The Society recommends a system of dynamic consent, that is, "collecting a first informed consent (IC) about SFs during the initial medical procedure motivating the primary genetic analysis. A second IC form is offered after the announcement of the primary findings so that the patient could, with more autonomy, differentiate the issues and confirm or refuse access to this information. The need to dissociate the announcement of primary findings from SFs was supported by patient associations to limit the psychological impact generated."

ETHICAL EXPLORATION

Introduction: screening criteria

Screening entails the offer of medical testing on the initiative of public health agencies or medical practitioners to people who do not yet have symptoms of illness or other reasons for seeking medical help with regard to the conditions targeted by the screening test. From an ethical point of view, this marks an important difference with medical testing on indication, for example, to find the cause for the patient's complaint. Because screening is by definition offered to those without an indication, and because all medical procedures may also have adverse effects, it is less obvious that the testing will on balance be beneficial. As famously stated by Sir Muir Gray and colleagues, "all screening programmes do harm; some do good as well, and, of these, some do more good than harm at reasonable cost" (Gray, Patnick, & Blanks, 2008). Apart from opportunity costs, adverse effects of screening include psychosocial sequelae and possible iatrogenic harm as a result of false-positive test results or overdiagnosis, as well as false reassurance caused by a false-negative test.

Conditions for responsible screening were formulated for the World Health Organization by Wilson and Jungner in the 1960s (Wilson & Jungner, 1968), and later refined and adapted by various groups of authors and organizations, particularly with a view to developments in genetic and reproductive screening (Health Council of the Netherlands, 2008). The core set of these conditions can be summarized in terms of:
— Proportionality requirements: it must be clearly established that early detection of the conditions screened for can lead to a significant reduction in the burden of disease in the target group in question, or to other outcomes useful to the participants; these advantages must clearly outweigh the disadvantages that screening can always have (for themselves or for others).
— Autonomy requirements: participation in screening and undergoing follow-up testing must be based on an informed and free choice.

— Justice requirements: the use of available healthcare resources in connection with and because of the program must be clearly shown to be acceptable in terms of cost-effectiveness and justice. Measures of prevention and therapeutic interventions indicated for those with a positive (nonfavorable) final diagnosis should be accessible to them.
— Quality requirements: laboratory and test procedures, but also other elements of the screening process, including counseling, consent, and monitoring procedures, should be evidence-based and state of the art.

Does nonreproductive OGS generally and do the ACMG's and SFMPP's proposals in particular meet these criteria? In our reflections below we will focus on the material conditions of proportionality, respect for autonomy, and justice. We will first address OGS in adults and secondly OGS in children (or minors).

OGS in competent adult patients
Proportionality
How should the requirement of a positive balance of benefits over harms be operationalized? Who is to be regarded as the recipient or recipients (benefits and harms for whom?), what possible benefits and harms should we think of, and how should these be balanced?

First, benefits and harms for whom? We should of course primarily think here of those to whom the screening is offered: people in the target group who may decide to have themselves screened. It is a matter for debate if possible benefits and harms for others (third parties) should be counted as well. In the context of reproductive screening it is accepted that, to some extent, third party harms do matter—think of the so-called "disability rights critique" (DRC) (Parens & Asch, 1999). Regardless of whether the DRC is convincing or not, most commentators agree that the impact of prenatal screening on the interests of people with relevant disabilities is a relevant consideration. And what about possible third-party benefits? As we will see, these may be relevant as well. For the moment, we focus on the possible benefits and harms for the members of the target group (the testees) themselves.

Second, what are the possible benefits and harms of (nonreproductive) OGS for the testees? Some of these may be more general, linked with such OGS per se, while others may be more specific, linked with OGS as proposed by the ACMG and the SFMPP.

Possible benefits
Possible benefits are primarily medical, more specifically: the primary and/or secondary prevention of serious hereditary oncological and cardiovascular disease (the SFMPP targets the first category only, the ACMG targets both), as well as of related suffering and premature death, first of all in the testees themselves, and secondly also in blood

relatives of testees with a "positive" test result. Clearly, these possible benefits may be considerable, assuming that the test quality also in terms of its clinical validity (the predictive value of positive findings) is high, that the preventive actions (including prophylactic surgery) to be taken by carriers are effective, and that access to these interventions will be guaranteed. Whether the latter condition will be met, clearly depends on the organization and funding of the relevant healthcare system—and will therefore be highly contextual. What should be regarded as a real benefit of OGS for detected carriers in some countries or states, may be just a theoretical benefit for those with similar findings in countries or states with other healthcare systems.

Clearly, possible benefits of OGS not only include health gains for the testees or their relatives, but also possible reproductive benefits, in terms of the avoidance (by testees or their relatives) of transmitting high risks of serious genetic disorders to future children. It is remarkable that benefits of the latter type are not explicitly addressed in the American and French Recommendations.

The ACMG estimates that 1% of sequencing reports will include an "incidental" or "secondary" variant from the proposed list (Green et al., 2013). This may be an underestimation; Leslie Biesecker (cf. Chapter 2: Genome Sequencing and Individual Responses to Results) estimates that "on average, about 2% of the people who undergo exome or genome sequencing will have a secondary finding variant," while a recent study in the Netherlands suggests that at least 1 in 39 individuals (2.6%) would be shown to carry an increased risk for a severe dominant disease if healthy individuals would be routinely screened for mutations in the ACMG minimum list of genes (Yntema et al., 2018).

Adaptations might lead to a further increase of possible benefits of OGS. First, as the ACMG states, an argument could be made for reporting a broader range of mutations predisposing for disease later in life. The ACMG decided to only include "unequivocally pathogenic mutations in genes where pathogenic variants lead to disease with very high probability." It is expected that more such mutations will be defined in the near future. Second, the ACMG's minimum list (so far) completely disregards pharmacogenomics—while some experts consider this to be "low-hanging fruit." Clearly, this needs further scrutiny. Inclusion of pharmacogenomics in the minimum list would mean that the type of possible medical benefits would be more diverse and include precision pharmacotherapy.

Risks

But what about the risks, that is, possible harms and disadvantages? These are highly diverse, and include medical, psychological, societal, and moral risks. Clearly, these types of risk are strongly interlinked and may partially overlap. Furthermore, like the possible benefits, some of the risks or possible harms may be of a more general nature, linked with OGS per se, while others may be more specific, linked with the content and conditions of OGS as proposed by the ACMG and the SFMPP.

First, the medical risks or concerns linked with medical and genetic assumptions and implications. These include (Burke et al., 2013; Holtzman, 2013):

— The lack of knowledge about the natural history of some of the diseases included in the minimum list;
— The lack of sufficient evidence that all mutations included in the minimum list are highly penetrant;
— The lack of sufficient evidence that the penetrance of the mutations included in the list is as high in the general population as in severely affected families (in the general population the predictive value of a positive result will tend to be lower than in high-risk contexts);
— Linked with the latter point, people who may wrongly assume that they are at very high risk may unnecessarily opt for invasive, risky, and irreversible interventions, such as prophylactic bilateral ovariectomy and mastectomy, etc.;
— The lack of sufficient evidence that the mutations included in the minimum list are all actionable;
— There is a lack of controlled studies of interventions.

According to the original (1.0) ACMG Recommendations (Green et al., 2013), applicants with an indication for clinical sequencing would not get access if they would refuse to accept the (complete) OGS package—a potentially grave medical harm and a grave moral harm, that is avoided in the revised version (2.0; see below).

The ACMG's procedure regarding the possible extension of the minimum list is not clear and raises questions (Wolf, Annas, & Elias, 2013). What about the ACMG's view that labs may look for additional genes, not included in the minimum list "as deemed appropriate" by themselves? And what about the individual doctor's discretionary competence? What material criteria and procedural safeguards should be in place and respected in order to forestall "wilder" OGS and to optimally meet the prerequisite of proportionality?

Robust procedures seem to become even more important now that it has been suggested recently to include *polygenic* risk prediction in clinical care (Khera et al., 2018). While it is key in public health to identify healthy individuals at high risk for actionable disease, the focus of genetic approaches so far is largely on finding carriers of rare monogenic mutations at several-fold increased risk. Although most disease risk is polygenic, it has not yet been possible to use polygenic predictors to identify individuals at risk levels comparable to those of monogenic mutations. However, researchers recently reported to have validated genome-wide polygenic scores (GPS) identifying individuals at greater than threefold increased risk for five common diseases: coronary artery disease, atrial fibrillation, type 2 diabetes, inflammatory bowel disease, and breast cancer. And this is just the beginning, so the authors state. There is a clear need for interdisciplinary debate about the pros and cons of possible screening for such GPS generally and about those of adding GPS to OGS in particular. Would proceeding in that direction be a good idea? Would it open the floodgates of medicalization? Or should we be realistic and acknowledge that "if we don't offer this, people will buy such tests elsewhere, to their own and our disadvantage …" (as suggested by a Dutch clinical geneticist in a recent discussion)?

Second, OGS may come with psychological concerns and challenges. People undergoing OGS will be confronted with risks of diseases they are unfamiliar with, at a time when they are trying to cope, deal, and give meaning to the totally different genetic problem for which clinical sequencing is indicated. How do people who have no family history of the diseases included in the OGS panel cope and deal with unexpected "positive" findings and preventive (including: reproductive) options? Do people feel empowered ("good to know, now I am in control and can take action!") or do they feel overwhelmed? What are barriers and facilitators for adequate support? What are patients' counseling needs, both with regard to possible direct health gains and in view of options for reproductive decision-making as enabled by OGS? Given the different setting, premature extrapolations from (mostly reassuring) psychological research in (non)carriers in affected families (cf. Chapter 5, Uncertainties in Genome Sequencing and Chapter 10, Genome Sequencing in Prenatal Testing and Screening: Lessons Learned From Broadening the Scope of Prenatal Genetics From Conventional Karyotyping to Whole Genome Microarray Analysis) should be avoided. What to think of the ACMG's argument that precisely in order to avoid overwhelming their patients, professionals should refrain from extensive pretest counseling aimed at clarifying testees' preferences (their further argument is that such counseling would be "impractical" also, because clinical sequencing will increasingly be ordered by clinicians with varying levels of ability and experience in genetic counseling).

Third, one should think of societal risks, particularly pertaining to possible adverse consequences for access to jobs and insurance—although there may be little evidence for the latter, especially when the diseases for which screenees are found to be at risk are preventable or treatable, as is the case with the diseases included in both the ACMG's and the SFMPP's minimum list (Joly, Saulnier, Osien, & Knoppers, 2014). These societal risks, however, are probably to a considerable degree contextual, given interjurisdiction variance.

Finally, there may be "moral risks" involved in OGS. This subtype of risk is linked to the moral framework set for OGS. As said, ACMG/OGS 2.0 avoids the iatrogenic moral harm linked with the original ACMG/OGS 1.0: people who refuse OGS are still entitled to have the indicated test for the condition that led them to seek medical help. But there may be two further moral risks: OGS as proposed in the ACMG 2.0 may still reflect insufficient respect for competent people's autonomy, and may provide insufficient protection of children against inappropriate testing for later-onset disorders (see below).

Taking account of the possible benefits and risks, what about the proportionality of OGS? The ACMG is convinced that their proposal—that is to say: both the offering of OGS per se and the recommended minimum list of genes or variants to be included in the screening package (the "content")—meets the criterion of proportionality, because the health benefits are large and the risks are minimal. In a later clarification document, the College stresses that not offering such OGS would be unethical and unprofessional (American College of Medical & Genomics, 2013), and McGuire et al. state in their supporting paper that the recommendations may count "as evidence of the standard of care" in the case of malpractice litigation (McGuire et al., 2013).

With regard to the claim that such OGS is proportional, we need to acknowledge that the proportionality prerequisite is inherently linked with what has been termed the "evidentiary model" (Wilfond & Nolan, 1993). At least for the moment, there are simply too many questions, unknowns, and uncertainties—that is to say, there is insufficient evidence—to justify the conclusion that the ACMG and SFMPP proposals meet the proportionality criterion. So the claim seems to be premature—which is not to say that such OGS is a priori unsound. If OGS is being offered, it should be considered research (Burke et al., 2013; Holtzman, 2013). That is to say that rigorous pilot studies investigating a well-described set of research questions should be organized (in line with relevant recommendations of screening authorities). The details of such studies need interdisciplinary discussion and are obviously beyond the scope of this normative reflection, as is the question of how such studies can be linked with the wider debate on rapid learning health care (Solomon & Bonham, 2013). A critical part of the evaluation of such pilot studies should be a reflection on adequate criteria for sufficient evidence (Schildmann et al., 2015).

If, as we suggest, the proportionality of the proposed OGS is indeed still open for debate and should be the subject of further research, the claim that such OGS would meet the standard of care is, a fortiori, premature.

Respect for autonomy

Respect for persons, mostly summarized as respect for autonomy, is a fundamental principle in modern ethics generally and health care in particular. IC is the means to show this respect (cf. also Chapter 3, Consenting Patients to Genome Sequencing).

The view that respect for autonomy is crucial, is, however, not self-evident. As Emanuel and Emanuel have shown in their seeding paper about different models of the patient—physician relationship, the traditional view is fundamentally paternalistic (Emanuel & Emanuel, 1992). In this model, the doctor acts as the patient's guardian, knowing and doing what is in the best interest of the patient. Assuming that doctors and patients share common, objective, criteria for what is best, selectively informing patients, in order to encourage the patient to do what the doctor suggests in the patients' best interest, is considered to be part of good clinical practice—"In the tension between the patient's autonomy and well-being, between choice and health, the paternalistic physician's emphasis is toward the latter" (Emanuel & Emanuel, 1992). While various modern models of the patient–physician relationship can be distinguished, they all reject this classic paternalism and aim—in various ways—at respecting "the patient as a person." As said earlier, this modern emphasis on respect for patient's autonomy is also being articulated in the normative framework of screening generally (cf. par. 4.1).

In the context of OGS, a cascade of decisions has to be made:

A. The decision whether or not to have oneself tested beyond the initial indication,

B. Deciding about the scope or content of the test, and

C. Deciding about the measure(s) to be taken if the test proves one to be a carrier, both with regard to the preservation of one' own health and with regard to informing relatives at high risk.

We focus the current reflection on the implications of respect for autonomy in the context of type A and type B decisions.

The decision whether or not to have oneself tested beyond the initial indication

The standard approach (as reflected in the screening criteria mentioned) would be (voluntary) IC, enabling the members of the target group to make an informed decision that meets their own values and preferences. This dominant view, which we accept, raises some questions for further reflection and research, including: what information has to be provided, how to avoid information overload, and what about the optimal timing of soliciting IC for OGS, given the fact that patients may be already overwhelmed by the problem for which they were seeking medical help in the first place? With regard to the latter, the French guidelines for OGS recommend a so-called dynamic consent (Pujol et al., 2018): involving a first stage of IC for the indicated test, followed by IC for the proposed OGS at a later stage—an example of *staging* IC, as recently discussed by several authors (Appelbaum et al., 2014; Bunnik, Janssens, & Schermer, 2013).

However, two alternatives for IC have been suggested in the consecutive ACMG Recommendations on OGS:

"All or nothing"—that is to say: a "coercive offer" (ACMG/OGS 1.0)

According to the original ACMG Recommendations, patients with an indication for clinical sequencing should not get access to such sequencing if they don't accept the recommended OGS. Clearly, this threat of sanctioning the patient's possible nonacceptance of OGS puts substantial pressure on his or her consent. In ethical terms, the provision of such conditional access to indicated clinical sequencing is a so-called "coercive offer." The main ACMG argument for this "all or nothing" view is that the doctor's and laboratory professionals' fiduciary duty to prevent harm trumps the principle of respect for autonomy, just as it does in the reporting of IF elsewhere in medical practice. As a consequence, "failing to report is unethical" (American College of Medical and Genomics, 2013).

This view is problematic for different reasons. First, with regard to the fiduciary duty of the medical doctor: the view that this duty overrules respect for autonomy disregards the extent to which the latter requirement codefines the content of the former: "In both ethics and law, the clinician has a core fiduciary duty to respect the patient's right to decide what testing to undergo ..." (Wolf et al., 2013). Besides, it is difficult to see how laboratory professionals also could have such a fiduciary duty, as they have no treatment relationship with the patient—and as a "shared fiduciary duty" would seriously undermine the professional autonomy of medical doctors as well.

Second, the ACMG claims that the search for and reporting of "IFs" of genomic sequencing is no different from what happens elsewhere in medical practice (think of the radiologist informing the patient about an unexpected tumor) and in clinical genetics in particular (think of the handling of "actionable" IFs). The analogy is, however, invalid, as it starts from the ACMG's problematic definition of IFs as "the results of a deliberate search" (Burke et al., 2013). While the examples from ordinary medicine and clinical genetics concern real IFs (i.e., the unavoidable byproducts of an indicated test), the ACMG's deliberately searched-for "IFs" are not. This clearly shows how "Sprachpolitik" (deliberate framing) may misguide ethical reflection and generate moral confusion and short-circuiting.

In line with other critics, we conclude that the original ACMG/OGS Recommendations are at odds with the principle of respect for autonomy and the right to self-determination of competent people, especially their right not to know (Wolf et al., 2013). In fact, the view that the doctor's fiduciary duty trumps autonomy amounts to a return to the classic, paternalistic doctor–patient relationship. It is ironical that this is being promoted in the context of implementing a technology commonly presented as an integral part of "PERSONalised medicine" (Burke et al., 2013). Moreover, adding to the irony, the promoters belong to a discipline that played a major, if not decisive, role in liberating medical practice from traditional paternalism. We share the concern that the original ACMG position may thus set a precedent for eroding the ethical foundations of modern medicine (cf. par. 4.2.1).

"Opt out" (ACMG/OGS 2.0)

In its "updated" Recommendations, the ACMG, taking account of its critics and of a survey administered by the ACMG to its membership in January 2014, argues that patients should be able to opt out. Although the ACMG still recommend: "At the time of testing, the patient should be made aware that, regardless of the specific indication for testing, laboratories will routinely analyze the sequence of a set of genes deemed to be highly actionable so as to detect pathogenic variants that may predispose to a severe but preventable outcome," the College now stipulates: "Patients should be informed during the consent process that, if desired, they may opt out of such analysis" (ACMG Board of Directors, 2015).

The "opt out" approach avoids the fundamental criticism regarding the coercive offer as entailed in the original proposal. The threat of sanctioning patients' nonacceptance of OGS is removed, thereby lifting a barrier for voluntary choice. Still, we question as to whether this opt-out approach sufficiently meets the principle of respect for autonomy. After all, presenting OGS as the default option still entails a significant, though more subtle, pressure to take part in OGS. Furthermore, voluntariness may be better guaranteed and facilitated by requiring an explicit authorization. Finally, the opt-out approach may not contribute to or even hinder a proper understanding by the patient of the goals and implications of OGS. After all, patients' deliberation is not encouraged.

We conclude that both alternatives suggested by the ACMG are debatable. To be more precise: the original "coercive offer" is simply unacceptable from an ethical and legal point of view, while the "opt-out" approach is second best. IC is required for OGS, as recommended by the French Society (Pujol et al., 2018).

Deciding about the scope or content of the test

Should people be given the option to have themselves screened for only part of the mutations included in the minimum list? Can the content of OGS be "personalized"? Starting from a "respect for autonomy" perspective, an affirmative answer seems to be the default position. The updated ACMG/OGS Recommendations, however, state: "At this time, given the practical concerns and inherent difficulty of counseling patients about the features of each disorder and every gene on an ever-changing list, it is not feasible for patients to be offered the option of choosing a subset of medically actionable genes for analysis. Thus, the decision regarding routine analysis should apply to the entire set of genes deemed actionable by the ACMG" (ACMG Board of Directors, 2015). Clearly, possible best practices that are not feasible, cannot be morally or legally obligatory. But the Recommendation just quoted simply entails a non sequitur: from the premise (probably correct) that it would be impractical to allow patients to develop a detailed personal OGS list, it does not follow that they should have no say at all regarding the scope of the test. The claim that *any* personalization is impractical lacks substantiation—and may, again, reflect traditional paternalism. We would argue that the ethical and practical "pros and cons" of personalization should be part of the proposed OGS pilot studies (cf. the Section on Proportionality). Relevant questions may include: what about the feasibility of allowing patients to have OGS just for mutations with KP, not for mutations with EP? What about patients who, for whatever personal reasons, want OGS only for relevant oncogenetic mutations, as is the focus of the SFMPP Recommendations, but not for cardiogenetic mutations? Or what if patients would ask to have only relevant pharmacogenetic variants included?

Whereas the SFMPP recommendations rightly insist that "the patient's autonomy and desire to know or to ignore SF results must be respected"(Pujol et al., 2018), their account of what they refer to as "dynamic consent" raises further questions. For instance, the recommendation that "Patients could decline at any time to be informed about these SFs even if they previously gave their approval" seems to entail that SFs that were actively screened for on the basis of the patient's initial consent, should not be returned if the patient changes his or her mind at the second stage. But in terms of the responsibilities of professionals, it is a different matter not to screen for certain genes out of respect for the patient's right not to know, and not to report findings of great relevance for the patient's own health or that of his or her close relatives (Dondorp, Sikkema-Raddatz, de Die-Smulders, & de Wert, 2012). Ideally one should try to avoid burdening professionals by generating health information that the patient does not want to receive. Although this cannot be completely avoided in the genomic era, an adequate IC procedure for OGS should try to minimize this problem as far as possible.

Justice

In their book on the *Ethics of Screening in Health Care and Medicine*, Juth and Munthe state that "if a screening programme is to have any chance of securing ethical defensibility in terms of health and autonomy, it can be expected to be quite expensive" (Juth & Munthe, 2012). If funded from public or collective resources, these costs must be justified in the context of the total healthcare budget. This is also a matter of "opportunity costs": introducing an expensive screening program might mean that other forms of screening, health interventions, or other public services cannot be paid for. As stressed by the Health Council of the Netherlands, the costs of screening must be defined not only as the cost of the (initial) screening test but also as including the costs of follow-up testing and ensuing interventions (Health Council of the Netherlands, 2008). The total should be calculated as net costs, that is, after deducting any savings as a result of timely prevention or treatment that was made possible by the screening.

In view of the costs of screening it is important to consider if alternative approaches would be more cost-effective. Especially where regarding genetic screening for individual risk assessment, the question arises whether collective measures (partly informed by recent epigenetics research) should not be favored, such as general health education, aimed at changing unhealthy lifestyles, or measures of health protection targeting the environment or the workplace (McGlone West, Blacksher, & Burke, 2017). We would like to emphasize the relevance of such collective measures, also for reasons of justice, and argue that such measures should be prioritized, even more so in poor-resource countries. But in more affluent countries, combinations of collective measures and screening programs aimed at individual risk assessment, more especially the identification of people at a high risk of serious-but-actionable disorders could be considered—if these programs meet the general criteria for responsible screening. What further alternatives for OGS may be considered in the light of justice? There are two possible alternatives for screening based on "opportunity."

Need versus opportunity

A first option would be to rather make use of the criterion of "need." So-called cascade testing (CT) focuses on mutations which are definitely of KP, highly penetrant, and actionable, but targets the relatives of a proband: persons who are known to be at an a priori high risk. In countries that have already implemented CT for a larger number of Mendelian onco- and cardiogenetic disorders (including a number of the disorders on the SFMPPs and ACMGs lists for OGS), the question is whether OGS has a clear added value in terms of identifying much more people with a comparably high genetic risk—which could justify the large additional resources needed; while in countries that did not yet implement CT for a larger number of Mendelian onco- and cardiogenetic disorders, the question is: what has priority from a distributive justice perspective: more CT, OGS, or a smart combination of both? The comparative analysis of these different approaches should pay due attention to the "opportunity" costs, taking account of scarce resources for health care.

Equity versus opportunity

Why limit screening for the mutations in the ACMG's minimum list to those who happen to have an indication for genomic testing? Why not offer universal instead of OGS, assuming that the risks and benefits for the screenees will be similar? Doesn't the opportunistic character of OGS entail an unequal access problem that is difficult to justify given the principle of formal justice: treat people differently only if there are good reasons for doing so? The ACMG's response to this is that in OGS, unlike separate population screening with its requirements of extensive and costly infrastructure, the patient has already presented to the medical care system, has been evaluated, and is under the care of a clinician. As much of the costs of the study have already been incurred for the primary indication, the cost-effectiveness ratio of OGS compares favorably to that of separate screening for the same conditions (Green et al., 2013). However, the difference may be smaller than suggested if OGS is performed on the basis of meaningful IC rather than a mere opt-out.

Our provisional conclusion is that OGS can be accepted for the time being, in the context of a research project. But if OGS would prove to be proportional (meaning that its benefits clearly outweigh the risks), justice considerations would seem to favor universal genomic screening, unless the ACMG's cost argument just mentioned is convincing. But even then, part of the OGS package might be offered universally, for instance, the pharmaco part, which would probably not involve extensive cost and infrastructure requirements.

Finally, if OGS is proportional and justified, the demarcation of the target group requires further reflection. Persons whose DNA is being sequenced in the context of preconception, prenatal, or neonatal genetic screening are not included in the ACMG's proposal. One might question the argument for this limitation. The SFMPP recommendations are explicitly limited to adults, but do not contain a further specification of the target group (Pujol et al., 2018). Here again, this is relevant for determining the proportionality of the proposal.

OGS in children

Traditional guidelines with regard to predictive testing in minors state that decisions about such testing should protect the best interests of the child.

In general, such testing should be postponed, for a combination of two moral reasons: the child's right to informational self-determination (or a so-called "open future") (a deontological argument), and the risk of harming the child (a consequentialist argument) (Borry et al., 2009; Clarke, 1998, 2010; De Wert, 1999). Obviously, there may be exceptions to the rule of postponement, when predictive testing serves an imminent medical interest of the child. There is a strong consensus that parents have decision-making authority: it is generally up to them to decide about predictive testing in children when this testing would meet accepted guidelines. In exceptional cases, however, where postponement of a predictive test might seriously endanger the health prospects of a child (think of a high risk carrying a predisposition to MEN type 2A), the overruling of a parental refusal of such testing might be morally justified (De Wert, 1999).

What about OGS in children? Like with regard to OGS in adults, the criteria of proportionality, respect for autonomy, and justice need scrutiny. Let us assume, for the moment, that OGS in children would meet the criteria of proportionality and justice—what, then, about the relevance and implications of respect for autonomy?

Mutatis mutandis, the distinction made earlier between type A, type B, and type C autonomy issues (cf. the Section 'OGS in competent adult patients') is relevant here as well:

A. The decision whether or not to have oneself tested beyond the initial indication;
B. Deciding about the scope or content of the screening test; and
C. Deciding about the measure(s) to be taken if the test proves the child to be a carrier.

Like before, we focus on A and B. Clearly, the implications of respect for autonomy are more complex here than in the context of OGS in adults. After all, OGS in incompetent children would involve not just the autonomy of the parents: their actual parental decision-making authority, but also the possible future autonomy of the child (but see below).

A. The decision whether or not to accept have one's child tested beyond the initial indication

If OGS in children would prove to be proportionate, the dominant framework would prescribe that parents have decision-making authority, more particularly that they are entitled to give or withhold IC. Here again, however, the ACMG takes a different view. While the original ACMG/OGS 1.0 Recommendations state that the indicated test in the child plus the proposed OGS should be offered as a package deal, putting strong pressure on parents to accept OGS, the updated ACMG/OGS 2.0 version recommends that "Parents should have the option during the consent process to opt out of such analysis"(ACMG Board of Directors, 2015). While the elimination of the former "coercive offer" is an important step, this is not sufficient: true parental IC is a necessary condition. That is not to say, however, that it is also a sufficient condition. To understand this, the content of the proposed OGS in children requires further scrutiny.

B. The scope or content of OGS in children

As expressed in, for example, the ESHG Recommendations regarding presymptomatic testing in children (European Society of Human Genetics, 2009) and the Dutch Health Council's framework for NGS (Dondorp & De Wert, 2010), it is widely felt that OGS in children should be questioned if this involves presymptomatic testing for late-onset (though preventable and/or treatable) conditions, like in both the ACMG 1.0 and 2.0 Recommendations.

The ACMG, however, does not agree with this, for two reasons.

First, so the ACMG argues, the traditional Recommendations are inconsistent with the general practice of respecting parental decision-making about their children's health. This seems to be a rather weak rebuttal, however, as parental authority is widely accepted to be conditional, not absolute.

Secondly, as suggested by McGuire et al. (2013): there would be no contradiction between the traditional, restrictive, guidelines regarding predictive testing in children for later-onset disorders and the ACMG Recommendations to engage in OGS for the same later-onset disorders in children *as the contexts are different*:

— In the traditional context of presymptomatic testing in relatives of a proband, the a priori risk of these relatives is *known*, so postponing presymptomatic testing in children for the disease at hand does not generate any harm—other relatives can simply ask for the test themselves;

— In the OGS context, however, there is *no known* risk of carrying a mutation included in the screening test. If a child is tested on indication for disease X, there are basically two options: (1) seek and report actionable SFs, running the risk of imposing information that offers potential medical benefit to both the child and relatives at risk; (2) do not seek and report actionable SFs, and as a consequence do not generate (or generate but withhold) information that is potentially life-saving for the child, one of its parents, and potentially other family members—information that they would otherwise have no reason to suspect. In practice, there will often be just one single opportunity to identify children carrying actionable mutations, so not taking advantage of this opportunity risks avoidable harm to children and their relatives.

Following the traditional view, this argument is not convincing, as it amounts to screening children "just to benefit others." But this categorical rebuttal may be too easy. If (a) predictive testing of children for truly actionable mutations for later-onset disorders really does not harm the child, and (b) it would prove to be the case that most children would opt for such testing later in life anyhow, and (c) such OGS in children would indeed sometimes, if not regularly, prevent serious harm in a parent or in relatives, is such OGS, then, clearly morally unsound? Or should we say that such considerations would strengthen the argument for *universal* screening of adults for the mutations on the list, on the condition that this would meet the criteria of proportionality and justice? The ACMG argues that the proposed OGS is the best possible option in this "transitional period," where people (parents) do not have easy access to inexpensive, readily interpretable sequencing in order to obtain personal risk information for the conditions enlisted (Green et al., 2013; McGuire et al., 2013). But, how, then, to demarcate this transitional period? And does this argument, if relevant, also apply to those regions and countries where access to genetic information, also in the context of cascade testing, is already well organized?

If we still want to extrapolate and apply the traditional recommendations regarding presymptomatic testing in children to OGS in children, it would be a non sequitur to conclude that any OGS in children is ethically problematic. After all, this depends upon the content of the proposed OGS. OGS in children for relevant pharmacogenetic variants and/or for early-onset disorders, like MEN type 2A, would probably fit the traditional normative framework. And last, but not least, it may be relevant to make a distinction between incompetent children that may or will probably become competent later in life on the one hand, and children that will (probably) not become competent later on the other. Clearly, arguments pertaining to children's future autonomy or "open future" are relevant for the first category only.

CONCLUSIONS

Is (nonreproductive) OGS to be regarded as ethically (and legally) acceptable or even recommendable, and if so, on what conditions? As said, according to the ESHG, genomic testing should be as targeted as possible, at least for the time being. This view may well be too strict—broader testing seems to be not a priori and altogether unsound, and might be justified, depending on the aim(s), content, context, and conditions of broader testing.

In the current normative exploration, we focused on the ACMG's OGS proposals on (nonreproductive) OGS, adding just a few comments on the recently published Recommendations of the SFMPP. In addressing the question whether OGS would meet the criteria for screening, we focused on the implications of the principles of proportionality, respect for autonomy, and justice.

With regard to the proportionality prerequisite, we argue that given the many uncertainties and concerns linked with OGS, it is premature to consider OGS for later-onset disorders to be proportional. Pilot studies are needed to investigate the opportunities and intricacies of OGS, taking account of different target groups, screening packages, contextual variables, and the views and experiences of relevant stakeholders, including patients. Such pilots are necessary for informed decision-making about OGS, the demarcation of the target group(s), the panel selection, and the provision of adequate information and counseling. Screening for pharmacogenomics needs separate scrutiny, as this may regard low-hanging fruit, raising relatively few difficult problems.

Some may consider OGS to be problematic because opportunity is an arbitrary inclusion criterion per se. They may argue that if predictive screening for highly penetrant, actionable mutations is offered at all, it should be offered either universally (because of equity) or in the context of cascade testing in affected families (because of need). But, also depending on the resources available for health care, OGS pilots may be justified at least as a temporary strategy, to generate data for a future, informed, debate about the possible proportionality of *universal* genomic screening for highly penetrant, actionable mutations, and/or as a complementary strategy, to be combined with *cascade* testing.

ACMG/OGS 1.0 entails a "coercive offer," which is at odds with both patients' autonomy rights and widely accepted criteria for screening. ACMG/OGS 2.0, recommending an "opt-out" approach, is a step in the right direction, but still fails to meet the IC standard, as recommended by the SFMPP. However, the dynamic consent approach proposed in the French Recommendations, also raises questions, as it may lead to burdening professionals with information that the patient on second thoughts does not want to receive, but that may be of great relevance for the patient's own health or that of his or her close relatives.

OGS in children for later-onset actionable mutations needs further ethical scrutiny—but there seem to be no valid principled objections to OGS in children for pharmacogenetic variants and early-onset actionable mutations (if such targeted OGS would meet the principles of proportionality and justice).

Finally, we need clear procedures and criteria for decision-making about possibly implementing OGS, the composition of the test panel, and its possible future expansion. The ACMG stimulates discussion with the medical community to explore the relevant issues "in the best interest of patients." Obviously, patients themselves should also have a say, they should not be reduced to the object of well-intended medical deliberations and interventions.

REFERENCES

ACMG Board of Directors. (2015). ACMG statement: Updated recommendations regarding analysis and reporting of secondary findings in clinical genome-scale sequencing. *Genetics in Medicine, 17*(1), 68–69.

American College of Medical., & Genomics, G. (2013). Incidental findings in clinical genomics: A clarification. *Genetics in Medicine, 15*(8), 664–666.

Appelbaum, P. S., Parens, E., Waldman, C. R., Klitzman, R., Fyer, A., Martinez, J., ... Chung, W. K. (2014). Models of consent to return of incidental findings in genomic research. *Hastings Center Report, 44*(4), 22–32.

Borry, P., Evers-Kiebooms, G., Cornel, M. C., Clarke, A., Dierickx, K., ... Professional Polic Committee (PPPC) of the European Society of Human Genetics (ESHG). (2009). Genetic testing in asymptomatic minors: Background considerations towards ESHG Recommendations. *European Journal of Human Genetics, 17*(6), 711–719.

Bunnik, E. M., Janssens, A. C., & Schermer, M. H. (2013). A tiered-layered-staged model for informed consent in personal genome testing. *European Journal of Human Genetics, 21*(6), 596–601.

Burke, W., Antommaria, A. H., Bennett, R., Botkin, J., Clayton, E. W., Henderson, G. E., ... Zimmern, R. (2013). Recommendations for returning genomic incidental findings? We need to talk!. *Genetics in Medicine, 15*(11), 854–859.

Clarke, A. (1998). *The genetic testing of children.* Oxford: BIOS Scientific Publishers.

Clarke, A. (2010). What is at stake in the predictive genetic testing of children? *Familial Cancer, 9*(1), 19–22.

De Wert, G. (1999). *Met het oog op de toekomst. Voortplantingstechnologie, erfelijkheidsonderzoek en ethiek.* Amsterdam: Thela Thesis.

Dondorp, W., & De Wert, G. (2010). *The thousand dollar genome: An ethical exploration.* The Hague: Health Council of the Netherlands.

Dondorp, W., Sikkema-Raddatz, B., de Die-Smulders, C., & de Wert, G. (2012). Arrays in postnatal and prenatal diagnosis: An exploration of the ethics of consent. *Human Mutation, 33*(6), 916–922.

Emanuel, E. J., & Emanuel, L. L. (1992). Four models of the physician-patient relationship. *JAMA: The Journal of the American Medical Association., 267*(16), 2221–2226.

European Society of Human Genetics. (2009). Genetic testing in asymptomatic minors: Recommendations of the European Society of Human Genetics. *European Journal of Human Genetics, 17*(6), 720–721.

Gray, J. A., Patnick, J., & Blanks, R. G. (2008). Maximising benefit and minimising harm of screening. *BMJ, 336*(7642), 480–483.

Green, R. C., Berg, J. S., Grody, W. W., Kalia, S. S., Korf, B. R., Martin, C. L., ... American College of Medical Genetics and Genomics. (2013). ACMG recommendations for reporting of incidental findings in clinical exome and genome sequencing. *Genetics in Medicine, 15*(7), 565–574.

Health Council of the Netherlands (2008). Screening: Between hope and hype. The Hague.

Holtzman, N. A. (2013). ACMG recommendations on incidental findings are flawed scientifically and ethically. *Genetics in Medicine, 15*(9), 750–751.

Joly, Y., Saulnier, K. M., Osien, G., & Knoppers, B. M. (2014). The ethical framing of personalized medicine. *Current Opinion in Allergy and Clinical Immunology, 14*(5), 404–408.

Juth, N., & Munthe, C. (2012). *The ethics of screening in health care and medicine: Serving society or serving the patient?* Dordrecht: Springer.

Khera, A. V., Chaffin, M., Aragam, K. G., Haas, M. E., Roselli, C., Choi, S. H., ... Kathiresan, S. (2018). Genome-wide polygenic scores for common diseases identify individuals with risk equivalent to monogenic mutations. *Nature Genetics, 50*(9), 1219–1224.

McGlone West, K., Blacksher, E., & Burke, W. (2017). Genomics, health disparities, and missed opportunities for the nation's research agenda. *JAMA: The Journal of the American Medical Association, 317* (18), 1831–1832.

McGuire, A. L., Joffe, S., Koenig, B. A., Biesecker, B. B., McCullough, L. B., Blumenthal-Barby, J. S., ... Green, R. C. (2013). Point-counterpoint. Ethics and genomic incidental findings. *Science, 340* (6136), 1047–1048.

Parens, E., & Asch, A. (1999). The disability rights critique of prenatal genetic testing. Reflections and recommendations. *Hastings Center Report, 29*(5), S1–S22.

Pujol, P., Vande Perre, P., Faivre, L., Sanlaville, D., Corsini, C., Baertschi, B., ... Geneviève, D. (2018). Guidelines for reporting secondary findings of genome sequencing in cancer genes: The SFMPP recommendations. *European Journal of Human Genetics, 26*(12), 1732–1742.

Schildmann, J., Boettcher, M., Gabriel, M., Ganser, A., Gottwald, S., Hessel, F., et al. (2015). 'Personalised medicine': Multidisciplinary perspectives and interdisciplinary recommendations on a framework for future research and practice. In J. Vollmann, V. Sandow, S. Waescher, & J. Schildmann (Eds.), *The ethics of personalised medicine: Critical perspectives* (pp. 259–280). Farnham: Ashgate/Routledge.

Schleidgen, S., Klingler, C., Bertram, T., Rogowski, W. H., & Marckmann, G. (2015). What is personalized medicine? Sharpening a vague term based on a systematic literature review. In J. Vollmann, V. Sandow, S. Waescher, & J. Schildmann (Eds.), *The ethics of personalised medicine: Critical perspectives* (pp. 9–24). Farnham: Ashgate/Routledge.

Solomon, M. Z., & Bonham, A. C. (2013). Ethical oversight of research on patient care. *Hastings Center Report,* S2–S3, Spec No.

van El, C. G., Cornel, M. C., Borry, P., Hastings, R. J., Fellmann, F., Hodgson, S. V., ... ESHG Public and Professional Policy Committee. (2013). Whole-genome sequencing in health care: recommendations of the European Society of Human Genetics. *European Journal of Human Genetics, 21*(6), 580–584.

Vollmann, J. (2015). Priority setting and opportunity costs in European public health care systems. In J. Vollmann, V. Sandow, S. Waescher, & J. Schildmann (Eds.), *The ethics of personalised medicine: Critical perspectives* (pp. 245–256). Farnham: Ashgate/Routledge.

Wilfond, B. S., & Nolan, K. (1993). National policy development for the clinical application of genetic diagnostic technologies. Lessons from cystic fibrosis. *JAMA: The Journal of the American Medical Association, 270*(24), 2948–2954.

Wilson, J. M. G., & Jungner, G. (1968). *Principles and practice of screening for disease.* Geneva: World Health Organization (WHO), Contract No: 34.

Woehlke, S., Perry, J., & Schicktanz, S. (2015). Taking it personally: Patients perspectives on personalised medicine and its ethical relevance. In J. Vollmann, V. Sandow, S. Waescher, & J. Schildmann (Eds.), *The ethics of personalised medicine: Critical perspectives* (pp. 129–147). Farnham: Ashgate/Routledge.

Wolf, S. M., Annas, G. J., & Elias, S. (2013). Point-counterpoint. Patient autonomy and incidental findings in clinical genomics. *Science, 340*(6136), 1049–1050.

Yntema H.G., Haer-Wigman L., Vulto-van Silfhout A., van der Schoot V., Gilissen C., Brunner H.G., et al. 1 in 39 individuals carries a dominant high-risk disease allele. Oral Presentation. Annual Congress European Society of Human Genetics Milan, Italy 18 June 2018.

CHAPTER 13

Summary of Key Areas for Research

Barbara B. Biesecker[1] and Aad Tibben[2]
[1]Distinguished Fellow, RTI International, Bethesda, MD, United States
[2]Department of Clinical Genetics, Leiden University Medical Centre, Leiden, The Netherlands

Genome sequencing has entered mainstream medical practice and is becoming more widely available by direct-to-consumer testing. Its use is primarily to diagnose rare conditions, to improve the accuracy of disease risk prediction and to identify the etiology of developmental disorders, such as autism with other features. Patients, consumers, and healthcare providers are increasingly more familiar with and interested in the contribution of genomics to health and diseases (Middleton et al., 2016). Studies of early adopters of sequencing demonstrate their interest in a variety of clinical results and often find personal utility from their results or results on their children (Facio et al., 2011; Sapp et al., 2014; van der Steen et al., 2018). Evidence to date, as reviewed in chapters of this book, has shown that receiving results of genome sequencing is appreciated by recipients with little evidence of psychological distress. As we have learned from medicine generally and from predictive and diagnostic single-gene testing, receiving information about a medical condition or hereditary risk from genome sequencing can often be better than the lack of an etiology, even when it does not (yet) guide treatment (Madeo, O'Brien, Bernhardt, & Biesecker, 2012; Yanes, Humphreys, McInerney-Leo, & Biesecker, 2017). In the context of cancer, genome sequencing has led to targeted and effective drugs for rare types of cancer at an astounding rate and patients' quality of life and life expectancy are improving with targeted treatments. Yet, the challenges of meeting end-users' expectations, ensuring accurate understanding of a variety of types of results, and helping recipients to manage receipt of uncertain results, argue for the importance of integrating genomics into an array of psychological considerations.

Within this volume, contributors have reported on responses to return of results in clinical and research populations. Expectations of end-users have been shown to depend on the consent process, understanding of the various types of results and their consequences, and the reason for pursuing testing. The complexities of understanding results, perceiving risks, and making informed healthcare decisions, argue for the importance of genetic counseling. Research is needed to further understand these clinical processes and sequencing outcomes and is greatly anticipated during this translation

Clinical Genome Sequencing
DOI: https://doi.org/10.1016/B978-0-12-813335-4.00013-1

of genomics into clinical care. The specific outcomes experienced by patients and consumers need to be well understood to maximize benefits from this new technology.

A collection of perspectives is reflected in this book that will help to inform healthcare providers in medicine and psychology to broaden their understanding of genome sequencing. While outcomes of initial studies that returned results are promising and reinforce the resilience of patients, studies of broader, more diverse populations and longitudinal studies are needed to appreciate health outcomes of translating genomics into care. Given the wide horizon of genome sequencing for clinical care, it is imperative to consider the priorities for psychological and social research relevant to maximizing the utility of genome sequencing for health benefits and to minimize potential adverse consequences. Psychological and social research should assess the practices and needs of healthcare professionals who provide genomics sequencing information, and the patients and consumers who are offered and receive results. On one hand, the quality and quantity of information conveyed, the quality of provider—patient communication, access to healthcare resources, and the consent process, need to be properly assessed to generate evidence to inform health policies and resource allocation. On the other hand, patients' preferences for information, level of understanding of sequencing results, psychological responses to results, management of uncertainty and secondary findings, use of recommended healthcare follow-up, access to and utilization of healthcare resources, and satisfaction and/or regret in the longer term require the further attention of psychological researchers (Gray et al., 2014; Khan et al., 2015; Robinson et al., 2018).

THE HEALTHCARE PROVIDER'S PERSPECTIVE

Genome sequencing is a major challenge for healthcare professionals. Whereas genetic counseling in the first decades of single-gene testing was the playing field and privilege of trained clinical geneticists and genetic counselors, nowadays medical specialists and other healthcare providers have or will become increasingly familiar with genetics and genomics. The technologies have only recently entered medicine and many healthcare professionals are not familiar with genetics concepts and implications for health to accurately inform their patients in an accessible way, and to help them make "good" decisions to undergo testing and to act on the results. While the explanation of a single mutation requires time and expertise, it is an even greater challenge to inform people in clear, comprehensible ways about the complexities of genome sequencing. These concern the potential for identification of an etiology, and the likely uncertainty or unreliability of some results. Healthcare professionals are trained to provide information tailored to their patients' existing knowledge and health literacy skills. In the case of uncertainty, eager providers may be tempted to defuse the patient's uncertainty and fear by conveying an abundance of information from which the patient absorbs little. When informing patients about the results of genome sequencing, healthcare professionals will require skills that enable them to manage the uncertainties and affective responses that the information may elicit in the patient. Clinical geneticists and genetic counselors are specially trained to share and discuss with their patients the complex information that genetic or genome diagnostics provide. They currently apply these skills in the context of genetics as outlined in contemporary practice definitions (Resta, 2006; Skirton & Patch, 2009). In genomics, it remains a priority that end-users "understand and adapt to the medical, psychological, and familial implications of genetic contributions to disease." It is to be expected that in the coming decades most healthcare providers will integrate use of genomic information into their practices. As such, they will be involved in providing information about genome sequencing and in returning results and making related healthcare recommendations. This raises the question of what healthcare professionals need to master about genome sequencing. How can they best be prepared to meet patient information needs for understanding the consequences of undergoing genome sequencing to facilitate informed decision-making? And how do we educate a majority of healthcare providers as the need for using genomics fans out across subspecialties? Perhaps we can expect that the next generation of young healthcare professionals will be familiar with the principles and scope of genomic sequencing (Austin, Semaka, & Hadjipavlou, 2014; Salari, Karczewski, Hudgins, & Ormond, 2013). Today we find ourselves in a period of transition which means that the current generation of healthcare providers needs education and resources to keep pace with a rapidly changing medical landscape. The decreasing costs make genome sequencing more accessible to the masses. Patients and consumers may have expectations that healthcare professionals cannot always meet, and they place high demands on the communication skills of professionals. According to Kessler, teaching the counseling model with attention to contextual, psychological, and informational dimensions may equip professionals with the appropriate skills to effectively communicate the results of genome sequencing (Kessler, 1997). Clearly, the weight of professional involvement depends on the weight of the results to be disclosed. It may be unrealistic to expect that all patients will receive their results in person. In the near future, patients may be informed and counseled using the telephone, video counseling, and online interactive platforms. Evidence suggests that these means may be as effective as in person return of results (Athens et al., 2017; Biesecker et al., 2018). Educational programs available online may increasingly provide basic education about what testing is available and what should be considered in making a decision. Examples of these programs currently exist and can be found online, developed by genetic testing companies. Although aimed at selling testing, some offer essential pretest information for consideration by potential users.

It is a challenge to determine the major needs of end-users to adequately consent them to genome sequencing. Studies are needed to determine what people minimally need to know to be able to give informed consent. Research is needed to establish what constitutes essential information. There is evidence that shorter, lower literacy consent information is as effective in educating participants in the potential benefits and harms of genome sequencing as longer, higher literacy consent information (Turbitt et al., 2018). Yet further studies are needed. From a psychological perspective, it is important that the patient is responsible for the choice, and that the healthcare professional does not impose a preferred option on the patient, unless sequencing may lead to a life-saving treatment as in the case of rare cancers in which case it may be medically recommended. Future studies should investigate the real-time communication process between healthcare providers and patients when discussing the outcome of genome sequencing to better understand which provider characteristics contribute to the patient's subsequent decision-making and whether they have facilitated a preference-based decision.

THE PATIENT'S OR CONSUMER'S PERSPECTIVE

Similar to healthcare professionals, patients and consumers are increasingly aware of genome sequencing. Genetics and genomics have entered daily news broadcasts and social conversation. People have expectations for the power of genomics, and their expectations are often unrealistic. In this era of empowerment, patients are aware that they have ownership over their body, including their genome. In a study by Gollust et al. (2012), few participants expressed deterministic views of genetic risk, suggesting that although expectations are high, they do not seem to be fatalistic. Yet, there remain debates about whether patients and consumers have a right to the return and interpretation of all types of information in their genome. Studies have shown a broad interest in results from genome sequencing, most strongly in receiving actionable information to offset health risks (Facio et al., 2013; Gollust et al., 2012; Middleton et al., 2016; Sanderson, Wardle, Jarvis, & Humphries, 2004). Yet the data are limited by the absence of large representative samples that are generalizable. Investigating more actual decision-making outcomes, rather than hypothetical choices or preferences prior to making a choice, will better predict future choices. We are still learning what characteristics leave some people more interested in learning about their health risks. We also do not know how likely it is that people will make sensible use of the information to pursue medical screening and to live a healthier lifestyle. Actionability is often equated with medical treatment such as prophylactic surgery or pharmacotherapy. However, experiences with single-gene testing have taught us all that actionability also includes relief from anxiety, gaining certainty, preparing for the future, facilitating end-of-life decisions, and learning reproductive options. As such, the actionability of results should be expanded to include psychological and moral factors that contribute to wellbeing. Moreover, differences in patient preferences for the return of results should be safeguarded.

From a cognitive perspective, research is needed on how people understand and process the information resulting from genome sequencing. Recent studies have shown that study participants who choose to learn results from genome sequencing have a reasonable understanding of the risk information (Gordon et al., 2012; van der Steen et al., 2018). Yet studies are needed to optimize the understanding of genomic risk information, and decision-making about options for following-up on results, and to facilitate patient adaptation to results. Decades of research into cognitive, affective, and motivating factors known to influence risk perception support the findings from single-gene testing that risk perception often deviates substantially from actual risk (Bonner, 2017; Vos et al., 2008; Vos, Oosterwijk, et al., 2011). Affective risk perception has been shown to be a stronger predictor of decisions and behavioral responses (Bonner, 2017; van Dooren et al., 2004). However, introduction of the TRIRISK model demonstrates that cognitive, affective, and experiential risk perceptions differ, and each predicts novel outcomes (Ferrer et al., 2018; Ferrer, Klein, Persoskie, Avishai-Yitshak, & Sheeran, 2016). These data suggest the need for further research in understanding risk perceptions of findings generated by genome sequencing. Experiences with single-gene counseling and testing have demonstrated the resilience and coping abilities of the vast majority of patients, suggesting that participants may self-select according to their confidence in their ability to manage the outcomes (Broadstock, Michie, & Marteau, 2000; Duisterhof, Trijsburg, Niermeijer, Roos, & Tibben, 2001; Paulsen et al., 2013). Most studies have also shown that small numbers of individuals are vulnerable to adverse outcomes. Psychological research has revealed factors that identify vulnerable patients, including pretest depression, psychiatric history, discordant relationship, lack of social support, and having children (Broadstock et al., 2000; Duisterhof et al., 2001; Paulsen et al., 2013).

Psychological studies in genome sequencing demonstrate that participants hold positive attitudes towards learning results (Bijlsma et al., 2018; Facio et al., 2013; Gray et al., 2016; Shahmirzadi et al., 2014; Taber, Klein, Ferrer, Han, et al., 2015; Taber, Klein, Ferrer, Lewis, et al., 2015). A positive attitude was correlated to the availability of treatment options, the perceived ambiguity of the return results, dispositional optimism, and self-efficacy. These variables need to be tested among other more diverse populations. There are likely differences in what factors influence people to prefer genome sequencing. Identification of factors that predict vulnerability among those pursuing genome sequencing is paramount for counseling and anticipating potential negative outcomes. Use of theoretical models to frame studies on how patients cope with health-threatening information, such as the Common Sense model (Cameron, Biesecker, Peters, Taber, & Klein, 2017), the Transactional Model of Stress and

Coping (TMSC), and the Perceived Uncertainty in Illness Theory (Mishel, 1988, 1990; Park & Folkman, 1997) can inform the selection of traits and variables previously shown to affect outcomes of the receipt of health-threatening information. Greater use of theory in psychological study design will enhance the quality of evidence generated and expedite evidence that is more likely to be generalizable.

Uncertainty, the subjective perception of ignorance (Han, Klein, & Arora, 2011), is at the heart of genome sequencing, as it is of life and of medicine generally. In a way, people need a certain, motivating level of uncertainty to feel and to take responsibility for their own health, their relationships, in particular with their offspring. People have different needs for certainty, and use different psychological strategies to cope with it, as was demonstrated by Vos and Bonner and their colleagues (Bonner, 2017; Vos et al., 2008; Vos, Oosterwijk et al., 2011). Healthcare providers who return results from genome sequencing provide a dose of uncertain results that may evoke a variety of emotions in patients. Strong emotions have the potential to hamper or block the processing of information. Consequently, the patient's uncertainty and confusion may increase, or decrease, when mental defense strategies are adopted. This may contribute to the common finding of inaccuracy in individual risk perceptions. Women with breast cancer, who had received appropriate pretest and post-test genetic counseling, after having received results from genetic testing for a mutation in the *BRCA1/2* genes were guided in their behavioral responses by their own perception of reality (Bonner, 2017; Vos, Oosterwijk et al., 2011; Vos, Stiggelbout et al., 2011). From a psychodynamic perspective, such violation of reality has meaning because adjustment of reality can enable the handling of emotions in response to that reality, even if only temporarily. Patients whose genome has been sequenced and who did not receive pretest counseling or appropriate clinical interpretation could needlessly undergo highly sensitive surveillance strategies (Hitch et al., 2014). Researchers and clinicians should help patients avoid this so-called "therapeutic misconception" in overestimating the clinical significance of findings (McGuire & Beskow, 2010).

A taxonomy of uncertainties in genome sequencing developed by Biesecker et al. (2017) categorizes uncertainty related to its sources, issues, and locus (Bloss et al., 2015; Han et al., 2011) and may serve to promote investigations into the dynamic nature of uncertainty. Further studies are needed to better understand variables that influence perceptions of uncertainty, recall, and reinterpretation of uncertain genetic information (Bonner, 2017).

After the return of results from genome sequencing, people are expected to follow up on recommendations for screening, treatment, or preventive activities. Whether it is a medical treatment option, healthy lifestyle, sharing of results with relatives, or reproductive choices, all require adjustment or change of behavior to optimize clinical utility. But we also know from a large number of studies that behavioral change is hard to achieve and maintain (Marteau & Lerman, 2001). It is important to ensure that individuals make medically and psychologically sound decisions. For example, studies of receipt of unexpected medically actionable secondary findings have found recipients to pursue recommended screening or health care and not excessive or unneeded care.

There is evidence of personal utility to end-users who value information for its own sake, even when there are no clinical recommendations that come from their results (Kohler, Turbitt, & Biesecker, 2017; Kohler, Turbitt, Lewis, et al., 2017). While genome sequencing will not be appropriated into health care based on personal utility alone, research participants outline numerous reasons they find value in learning information about themselves. There is overlap between personal and clinical utility. For example, participants place value on sharing health risk information with relatives. If relatives act on the information, this demonstrates clinical utility. Yet, research participants value having the opportunity to share risk information in their family, even if it is not acted on, demonstrating personal utility. Studies of the broad range of reasons participants have interest in undergoing genome sequencing are needed as they will help to predict outcomes of testing as well.

It is not uncommon that healthcare professionals witness their patients make decisions that give pause for thought, such as not informing adult children about carriers of a breast cancer or a Huntington gene mutation. Patients often imbue results with unique meaning. Yet even among those who adequately recall test results a deviant meaning can be construed. Studies by Vos and van Dooren showed that affective risk perceptions were a stronger predictor of decision-making than the actual risk communicated by the genetic counselor (van Dooren et al., 2004; Vos, Stiggelbout et al., 2011). Such data beg the question; how important is accurate understanding of risk given the complexity of results from genome sequencing? Distinct from single-gene testing, it will be impossible to interpret the majority of information generated from sequencing. The results that are returned will include variants of unknown significance, and variants with pleotrophic effects on health risks. These demand high literacy and numeracy (probability) to understand their potential consequences. Simplified explanations for complex information will be needed for patients to grasp basic concepts.

Recent studies suggest that direct-to-consumer genome testing does not lead to adverse psychological problems (Boeldt, Schork, Topol, & Bloss, 2015). Moreover, it may result in healthy behavioral changes, although it is acknowledged that there are individual differences in psychological outcomes following testing. Again, these first impressions in small groups should be further validated and evaluated in larger populations that include more diverse samples of individuals purchasing genetic testing for personal use.

In conclusion, the availability of genome sequencing can be regarded as a remarkable breakthrough in the changing landscape of health care, with opportunities for patients and consumers to take responsibility for their current and future health. The big challenge is to make sensible use of genomic information and healthcare resources. Psychological research has the potential to inform healthcare providers and patients about the best ways to convey and process information, and about how to adapt preferable behavior to optimize the benefits of genome sequencing. To this end, decision tools and behavioral interventions need to be further developed and tested to empower patients and consumers.

REFERENCES

Athens, B. A., Caldwell, S. L., Umstead, K. L., Connors, P. D., Brenna, E., & Biesecker, B. B. (2017). A systematic review of randomized controlled trials to assess outcomes of genetic counseling. *Journal of Genetic Counseling, 26*(5), 902–933. Available from https://doi.org/10.1007/s10897-017-0082-y.

Austin, J., Semaka, A., & Hadjipavlou, G. (2014). Conceptualizing genetic counseling as psychotherapy in the era of genomic medicine. *Journal of Genetic Counseling, 23*(6), 903–909. Available from https://doi.org/10.1007/s10897-014-9728-1.

Biesecker, B. B., Lewis, K. L., Umstead, K. L., Johnston, J. J., Turbitt, E., Fishler, K. P., . . . Biesecker, L. G. (2018). Web platform vs in-person genetic counselor for return of carrier results from exome sequencing: A randomized clinical trial. *JAMA Internal Medicine, 178*(3), 338–346. Available from https://doi.org/10.1001/jamainternmed.2017.8049.

Biesecker, B. B., Woolford, S. W., Klein, W. M. P., Brothers, K. B., Umstead, K. L., Lewis, K. L., & Han, P. K. J. (2017). PUGS: A novel scale to assess perceptions of uncertainties in genome sequencing. *Clin Genet, 92*(2), 172–179.

Bijlsma, R. M., Wessels, H., Wouters, R. H. P., May, A. M., Ausems, M., Voest, E. E., & Bredenoord, A. L. (2018). Cancer patients' intentions towards receiving unsolicited genetic information obtained using next-generation sequencing. *Familial Cancer, 17*(2), 309–316. Available from https://doi.org/10.1007/s10689-017-0033-7.

Bloss, C. S., Zeeland, A. A., Topol, S. E., Darst, B. F., Boeldt, D. L., Erikson, G. A., . . . Torkamani, A. (2015). A genome sequencing program for novel undiagnosed diseases. *Genetics in Medicine, 17*(12), 995–1001. Available from https://doi.org/10.1038/gim.2015.21.

Boeldt, D. L., Schork, N. J., Topol, E. J., & Bloss, C. S. (2015). Influence of individual differences in disease perception on consumer response to direct-to-consumer genomic testing. *Clinical Genetics, 87*(3), 225–232. Available from https://doi.org/10.1111/cge.12419.

Bonner, D. (2017). Understanding how clients make meaning of their variants of uncertain significance in the age of multi-gene panel testing for cancer susceptibility (MSc). John Hopkins University, Baltimore, MD.

Broadstock, M., Michie, S., & Marteau, T. (2000). Psychological consequences of predictive genetic testing: A systematic review. *European Journal of Human Genetics, 8*(10), 731–738. Available from https://doi.org/10.1038/sj.ejhg.5200532.

Cameron, L. D., Biesecker, B. B., Peters, E., Taber, J. M., & Klein, W. M. P. (2017). Self-regulation principles underlying risk perception and decision making within the context of genomic testing. *Social Personality Psychology Compass, 11*(5), e12315. Available from https://doi.org/10.1111/spc3.12315.

Duisterhof, M., Trijsburg, R. W., Niermeijer, M. F., Roos, R. A., & Tibben, A. (2001). Psychological studies in Huntington's disease: Making up the balance. *Journal of Medical Genetics, 38*(12), 852–861.

Facio, F. M., Brooks, S., Loewenstein, J., Green, S., Biesecker, L. G., & Biesecker, B. B. (2011). Motivators for participation in a whole-genome sequencing study: Implications for translational genomics research. *European Journal of Human Genetics, 19*(12), 1213–1217. Available from https://doi.org/10.1038/ejhg.2011.123.

Facio, F. M., Eidem, H., Fisher, T., Brooks, S., Linn, A., Kaphingst, K. A., ... Biesecker, B. B. (2013). Intentions to receive individual results from whole-genome sequencing among participants in the ClinSeq study. *European Journal of Human Genetics*, *21*(3), 261−265. Available from https://doi.org/10.1038/ejhg.2012.179.

Ferrer, R. A., Klein, W. M., Persoskie, A., Avishai-Yitshak, A., & Sheeran, P. (2016). The tripartite model of risk perception (TRIRISK): Distinguishing deliberative, affective, and experiential components of perceived risk. *Annals of Behavioral Medicine*, *50*(5), 653−663. Available from https://doi.org/10.1007/s12160-016-9790-z.

Ferrer, R. A., Klein, W. M. P., Avishai, A., Jones, K., Villegas, M., & Sheeran, P. (2018). When does risk perception predict protection motivation for health threats? A person-by-situation analysis. *PLoS One*, *13*(3), e0191994. Available from https://doi.org/10.1371/journal.pone.0191994.

Gollust, S. E., Gordon, E. S., Zayac, C., Griffin, G., Christman, M. F., Pyeritz, R. E., ... Bernhardt, B. A. (2012). Motivations and perceptions of early adopters of personalized genomics: Perspectives from research participants. *Public Health Genomics*, *15*(1), 22−30. Available from https://doi.org/10.1159/000327296.

Gordon, E. S., Griffin, G., Wawak, L., Pang, H., Gollust, S. E., & Bernhardt, B. A. (2012). "It's not like judgment day": Public understanding of and reactions to personalized genomic risk information. *Journal of Genetic Counseling*, *21*(3), 423−432. Available from https://doi.org/10.1007/s10897-011-9476-4.

Gray, S. W., Martins, Y., Feuerman, L. Z., Bernhardt, B. A., Biesecker, B. B., Christensen, K. D., ... CSER Consortium Outcomes and Measures Working Group. (2014). Social and behavioral research in genomic sequencing: Approaches from the Clinical Sequencing Exploratory Research Consortium Outcomes and Measures Working Group. *Genetics in Medicine*, *16*(10), 727−735. Available from https://doi.org/10.1038/gim.2014.26.

Gray, S. W., Park, E. R., Najita, J., Martins, Y., Traeger, L., Bair, E., ... Joffe, S. (2016). Oncologists' and cancer patients' views on whole-exome sequencing and incidental findings: Results from the CanSeq study. *Genet Med*, *18*(10), 1011−1019. Available from https://doi.org/10.1038/gim.2015.207.

Han, P. K., Klein, W. M., & Arora, N. K. (2011). Varieties of uncertainty in health care: A conceptual taxonomy. *Medical Decision Making*, *31*(6), 828−838.

Hitch, K., Joseph, G., Guiltinan, J., Kianmahd, J., Youngblom, J., & Blanco, A. (2014). Lynch syndrome patients' views of and preferences for return of results following whole exome sequencing. *Journal of Genetic Counseling*, *23*(4), 539−551. Available from https://doi.org/10.1007/s10897-014-9687-6.

Kessler, S. (1997). Psychological aspects of genetic counseling. IX. Teaching and counseling. *Journal of Genetic Counseling*, *6*(3), 287−295. Available from https://doi.org/10.1023/A:1025676205440.

Khan, C. M., Rini, C., Bernhardt, B. A., Roberts, J. S., Christensen, K. D., Evans, J. P., ... Henderson, G. E. (2015). How can psychological science inform research about genetic counseling for clinical genomic sequencing? *Journal of Genetic Counseling*, *24*(2), 193−204. Available from https://doi.org/10.1007/s10897-014-9804-6.

Kohler, J. N., Turbitt, E., & Biesecker, B. B. (2017). Personal utility in genomic testing: A systematic literature review. *European Journal of Human Genetics*, *25*(6), 662−668. Available from https://doi.org/10.1038/ejhg.2017.10.

Kohler, J. N., Turbitt, E., Lewis, K. L., Wilfond, B. S., Jamal, L., Peay, H. L., ... Biesecker, B. B. (2017). Defining personal utility in genomics: A Delphi study. *Clinical Genetics*, *92*(3), 290−297. Available from https://doi.org/10.1111/cge.12998.

Madeo, A. C., O'Brien, K. E., Bernhardt, B. A., & Biesecker, B. B. (2012). Factors associated with perceived uncertainty among parents of children with undiagnosed medical conditions. *American Journal of Medical Genetics Part A*, *158A*(8), 1877−1884. Available from https://doi.org/10.1002/ajmg.a.35425.

Marteau, T. M., & Lerman, C. (2001). Genetic risk and behavioural change. *BMJ*, *322*(7293), 1056−1059.

McGuire, A. L., & Beskow, L. M. (2010). Informed consent in genomics and genetic research. *Annual Review of Genomics and Human Genetics*, *11*, 361−381. Available from https://doi.org/10.1146/annurev-genom-082509-141711.

Middleton, A., Morley, K. I., Bragin, E., Firth, H. V., Hurles, M. E., Wright, C. F., . . . study, D. D. D. (2016). Attitudes of nearly 7000 health professionals, genomic researchers and publics toward the return of incidental results from sequencing research. *European Journal of Human Genetics*, *24*(1), 21−29. Available from https://doi.org/10.1038/ejhg.2015.58.

Mishel, M. H. (1988). Uncertainty in illness. *Image: The Journal of Nursing Scholarship*, *20*(4), 225−232.

Mishel, M. H. (1990). Reconceptualization of the uncertainty in illness theory. *Image: The Journal of Nursing Scholarship*, *22*(4), 256−262.

Park, C. L., & Folkman, S. (1997). Meaning in the context of stress and coping. *Review of General Psychology*, *1*(2), 115−144.

Paulsen, J. S., Nance, M., Kim, J. I., Carlozzi, N. E., Panegyres, P. K., Erwin, C., . . . Williams, J. K. (2013). A review of quality of life after predictive testing for and earlier identification of neurodegenerative diseases. *Progress in Neurobiology*, *110*, 2−28. Available from https://doi.org/10.1016/j.pneurobio.2013.08.003.

Resta, R. G. (2006). Defining and redefining the scope and goals of genetic counseling. *Am J Med Genet C Semin Med Genet*, *142C*(4), 269−275.

Robinson, J. O., Wynn, J., Biesecker, B. B., Brothers, K. B., Patrick, D. L., Rini, C., . . . Gray, S. W. (2018). A meta-analysis of psychological outcomes across 8 clinical sequencing exploratory research consortium studies. *Genetics in Medicine (under review)*.

Salari, K., Karczewski, K. J., Hudgins, L., & Ormond, K. E. (2013). Evidence that personal genome testing enhances student learning in a course on genomics and personalized medicine. *PLoS One*, *8*(7), e68853. Available from https://doi.org/10.1371/journal.pone.0068853.

Sanderson, S. C., Wardle, J., Jarvis, M. J., & Humphries, S. E. (2004). Public interest in genetic testing for susceptibility to heart disease and cancer: A population-based survey in the UK. *Preventive Medicine*, *39*(3), 458−464. Available from https://doi.org/10.1016/j.ypmed.2004.04.051.

Sapp, J. C., Dong, D., Stark, C., Ivey, L. E., Hooker, G., Biesecker, L. G., & Biesecker, B. B. (2014). Parental attitudes, values, and beliefs toward the return of results from exome sequencing in children. *Clinical Genetics*, *85*(2), 120−126. Available from https://doi.org/10.1111/cge.12254.

Shahmirzadi, L., Chao, E. C., Palmaer, E., Parra, M. C., Tang, S., & Gonzalez, K. D. (2014). Patient decisions for disclosure of secondary findings among the first 200 individuals undergoing clinical diagnostic exome sequencing. *Genetics in Medicine*, *16*(5), 395−399. Available from https://doi.org/10.1038/gim.2013.153.

Skirton, H., & Patch, C. (2009). *Genetics for the Health Sciences*. Bloxham: Scion Publishing Ltd.

Taber, J. M., Klein, W. M., Ferrer, R. A., Han, P. K., Lewis, K. L., Biesecker, L. G., & Biesecker, B. B. (2015). Perceived ambiguity as a barrier to intentions to learn genome sequencing results. *Journal of Behavioral Medicine*, *38*(5), 715−726. Available from https://doi.org/10.1007/s10865-015-9642-5.

Taber, J. M., Klein, W. M., Ferrer, R. A., Lewis, K. L., Biesecker, L. G., & Biesecker, B. B. (2015). Dispositional optimism and perceived risk interact to predict intentions to learn genome sequencing results. *Health Psychology*, *34*(7), 718−728. Available from https://doi.org/10.1037/hea0000159.

Turbitt, E., Chrysostomou, P. P., Peay, H. L., Heidlebaugh, A. R., Nelson, L. M., & Biesecker, B. B. (2018). A randomized controlled study of a consent intervention for participating in an NIH genome sequencing study. *Eur J Hum Genet*, *26*(5), 622−630.

van der Steen, S. L., Bunnik, E. M., Polak, M. G., Diderich, K. E. M., Verhagen-Visser, J., Govaerts, L. C. P., . . . Riedijk, S. R. (2018). Choosing between higher and lower resolution microarrays: Do pregnant women have sufficient knowledge to make informed choices consistent with their attitude? *Journal of Genetic Counseling*, *27*(1), 85−94. Available from https://doi.org/10.1007/s10897-017-0124-5.

van Dooren, S., Rijnsburger, A. J., Seynaeve, C., Duivenvoorden, H. J., Essink-Bot, M. L., Tilanus-Linthorst, M. M., . . . Tibben, A. (2004). Psychological distress in women at increased risk for breast cancer: The role of risk perception. *European Journal of Cancer*, *40*(14), 2056−2063. Available from https://doi.org/10.1016/j.ejca.2004.05.004.

Vos, J., Oosterwijk, J. C., Gomez-Garcia, E., Menko, F. H., Jansen, A. M., Stoel, R. D., . . . Stiggelbout, A. M. (2011). Perceiving cancer-risks and heredity-likelihood in genetic-counseling: How counselees recall and interpret BRCA 1/2-test results. *Clinical Genetics*, *79*(3), 207−218. Available from https://doi.org/10.1111/j.1399-0004.2010.01581.x.

Vos, J., Otten, W., van Asperen, C., Jansen, A., Menko, F., & Tibben, A. (2008). The counsellees' view of an unclassified variant in BRCA1/2: Recall, interpretation, and impact on life. *Psychooncology*, *17* (8), 822–830. Available from https://doi.org/10.1002/pon.1311.

Vos, J., Stiggelbout, A. M., Oosterwijk, J., Gomez-Garcia, E., Menko, F., Collee, J. M., ... Tibben, A. (2011). A counselee-oriented perspective on risk communication in genetic counseling: Explaining the inaccuracy of the counselees' risk perception shortly after BRCA1/2 test result disclosure. *Genetics in Medicine*, *13*(9), 800–811. Available from https://doi.org/10.1097/GIM.0b013e31821a36f9.

Yanes, T., Humphreys, L., McInerney-Leo, A., & Biesecker, B. (2017). Factors associated with parental adaptation to children with an undiagnosed medical condition. *Journal of Genetic Counseling*, *26*(4), 829–840. Available from https://doi.org/10.1007/s10897-016-0060-9.

INDEX

Note: Page numbers followed by "*t*" refer to tables.

Printed in the United States
By Bookmasters